SpringerBriefs in Applied Sciences and Technology

SpringerBriefs present concise summaries of cutting-edge research and practical applications across a wide spectrum of fields. Featuring compact volumes of 50 to 125 pages, the series covers a range of content from professional to academic.

Typical publications can be:

- A timely report of state-of-the art methods
- An introduction to or a manual for the application of mathematical or computer techniques
- A bridge between new research results, as published in journal articles
- A snapshot of a hot or emerging topic
- An in-depth case study
- A presentation of core concepts that students must understand in order to make independent contributions

SpringerBriefs are characterized by fast, global electronic dissemination, standard publishing contracts, standardized manuscript preparation and formatting guidelines, and expedited production schedules.

On the one hand, **SpringerBriefs in Applied Sciences and Technology** are devoted to the publication of fundamentals and applications within the different classical engineering disciplines as well as in interdisciplinary fields that recently emerged between these areas. On the other hand, as the boundary separating fundamental research and applied technology is more and more dissolving, this series is particularly open to trans-disciplinary topics between fundamental science and engineering.

Indexed by EI-Compendex, SCOPUS and Springerlink.

More information about this series at http://www.springer.com/series/8884

Rabiu Muazu Musa · Anwar P. P. Abdul Majeed ·
Muhammad Zuhaili Suhaimi ·
Mohd Azraai Mohd Razman ·
Mohamad Razali Abdullah ·
Noor Azuan Abu Osman

Machine Learning in Elite Volleyball

Integrating Performance Analysis,
Competition and Training Strategies

 Springer

Rabiu Muazu Musa
Universiti Malaysia Terengganu
Kuala Terengganu, Malaysia

Muhammad Zuhaili Suhaimi
Universiti Malaysia Terengganu
Kuala Terengganu, Malaysia

Mohamad Razali Abdullah
Faculty of Applied Social Science
Universiti Sultan Zainal Abidin
Kuala Terengganu, Malaysia

Anwar P. P. Abdul Majed
Faculty of Manufacturing and Mechatronic
Engineering Technology
Universiti Malaysia Pahang
Pekan, Pahang, Malaysia

Mohd Azraai Mohd Razman
Faculty of Manufacturing and Mechatronic
Engineering Technology
Universiti Malaysia Pahang
Pekan, Pahang, Malaysia

Noor Azuan Abu Osman
Faculty of Engineering
University of Malaya
Kuala Lumpur, Malaysia

ISSN 2191-530X   ISSN 2191-5318   (electronic)
SpringerBriefs in Applied Sciences and Technology
ISBN 978-981-16-3191-7   ISBN 978-981-16-3192-4   (eBook)
https://doi.org/10.1007/978-981-16-3192-4

This Springer imprint is published by the registered company Springer Nature Singapore Pte Ltd.
The registered company address is: 152 Beach Road, #21-01/04 Gateway East, Singapore 189721, Singapore

This book is dedicated to my beloved family, my precious wife Nafisa Usman Yahaya as well as my heroic son Musa Rabiu Muazu.

Rabiu Muazu Musa

I would like to dedicate this book to my family, P. P. Abdul Majeed K. Idros, Sulleha Eramu, Noriza P. P. Abdul Majeed, as well as my wife Sharifah Maszura Syed Mohsin, and not forgetting the apples of my eyes, Saffiya Anwar and Yusuff Anwar.

Anwar P. P. Abdul Majeed

This book is dedicated to all those who encouraged me to fly towards my dreams.

Muhammad Zuhaili Suhaimi

*Once for the watching eye
Twice for those dissatisfied
Three times they did nothing
Four times I was still standing*

Mohd Azraai Mohd Razmaan

I would like to dedicate this book to my wife, my daughter, and my son.

Mohamad Razali Abdullah

I would like to thank my wife Liana for her love. Thank you for being my best friend. I owe you everything.

Noor Azuan Abu Osman

Acknowledgements

We would like to acknowledge Prof. Dr. Zahari Taha for the guidance as well as the valuable suggestions for making the realization of this book possible. We would also like to extend our gratitude to the 2020 UMT Open Volleyball Championship Committee. We wish to extend our sincere gratitude to the management of the UMT Sports Centre for providing us with the support and the facilities to carry out this study. We would also wish to acknowledge the coaches as well as the managers of all the volleyball teams that participated in the tournament for their support throughout the data collection process.

Rabiu Muazu Musa
Anwar P. P. Abdul Majeed
Muhammad Zuhaili Suhaimi
Mohd Azraai Mohd Razman
Mohamad Razali Abdullah
Noor Azuan Abu Osman

Contents

Chapter 1
Nature of Volleyball Sport, Performance Analysis in Volleyball, and the Recent Advances of Machine Learning Application in Sports

Abstract This chapter presents the overview, nature, and history of the volleyball sport. The chapter also highlights the performance-related parameters that contribute to the successful delivery of performance in this sport. The recent advances in the application of various machine learning models towards solving the classification and regression problems associated with the data often acquired in the sporting domain are also provided. Moreover, the detailed procedures of the participants' recruitment, data collection techniques via performance analysis as well as various univariate statistical analyses employed to achieve the purpose of the present study have also been presented.

Keywords Volleyball · Machine learning models · Clustering algorithms · Univariate analysis · Performance analysis · Performance indicators

1.1 An Overview of Volleyball Sport

The sport of volleyball is considered as one of the famous sports played by various countries at varying level of expertise. Typically, the sport is considered as a team game that involves frequent short bout of high- and low-intensity jumping and strolling activity [1]. The sport of volleyball could be played both as competitive and for recreational purposes. Although, a different version of the sport is played around the globe, however, the sport has been widely adapted and played across different age groups and gender to fit several categories of players either with abilities or with disabilities. During the gameplay, a net typically divides the court in half, whilst each team that consist of five players defends the chosen area. For the team to earn points and successfully defend their selected territory, they are required to serve the ball across the net such that it lands on the opponent's side. As soon as the ball lands on the opponent's territory, the opposing team could then have a maximum of three touches on their selected side of the court before sending the ball back over the demarcated net. To earn a point, a team is required to ground the ball on the floor of the opponents' court or otherwise compel the opponent to either fail to return the ball or send the ball outside the selected area of the court. It is worth mentioning that the game of volleyball follows the rotational order of movement, unlike the other related games. In the volleyball game, the rotational system of the players on positions is compulsory in a clockwise manner to permit all the players on the court to have the opportunity of serving the ball.

The history of volleyball could be traced as far back as 1895 when it was first introduced by William Morgan. Despite the tremendous requirement of speed, agility, strength as well as power in the sport, the sport was initially invented to provide the players with a less violent experience as opposed to other games such as basketball or soccer. It is worth noting that the first name of the sport was 'mintonette' when it was initially introduced; nonetheless, considering the routine performance of volleying shots in a particular game, the name was later changed to its present name 'Volleyball'. Moreover, the nature of the game was quite different from the current format. It was noted that when the game was first introduced, the players were only required to alternately play the ball across the net and no limitation on the number of players allowed to play in a team during the competition. The major changes in the rules appeared in 1912 in which the formal guidelines were established that consisted of the number of players, the touches permitted as well as the rotational movement of the players after a point is won. The official governing body of the sport known as Federation International Volleyball (FIVB) was first established in 1947, and in 1964, the sport made its first Olympic debut at the Tokyo Olympic games for both male and female players [2]. Whilst there are two types of volleyball sports, i.e. beach volleyball and indoor volleyball, however, it is important to note that the present research exclusively focused on the indoor variation of the game, and therefore, when a volleyball is mentioned herein, we are particularly referring to the men indoor volleyball.

1.2 The Nature, Attributes, and Skill Requirements of the Volleyball Sport

The sport of volleyball is considered as an intermittent as well as a non-inversion game that necessitates the players to compete in an atmosphere that is characterized by short bouts of great intensity movement interwoven by a considerable low level of activity [3, 4]. Due to these high demands of physical fitness, the neuromuscular system of the players is often tasked with the need of high aerobic and anaerobic ability to facilitate performance during various activities such as sprints, jumps to block the opponent ball and during the execution of spiking as well as the repeated movement in the court during competition [5]. It can then be deduced that the players are expected to possess certain physical fitness-related elements such as endurance, speed upper, and lower body strength as well as power amongst others in tandem to the other team-based sports [6, 7]. Moreover, anthropometric-related variables have also been reported to be essential predictors of performance in volleyball [8, 9].

Since a successful performance in the game of volleyball is reliant upon a multitude of factors for a volleyball player to excel, many researchers have thus far attempted to apply various components to predict selection as well as identify talents within the sport [10–12]. For instance, previous researchers differentiated the volleyball players according to playing position and level of expertise [10]. It was established from the findings that the middle blockers were taller, more ectomorphic and endomorphic, whereas liberos were shorter, less ectomorphic, more mesomorphic, and more endomorphic than players in the other positions. More successful players in all positions had a lower body mass index, were less mesomorphic and endomorphic, and more ectomorphic than less successful players. Furthermore, more successful players showed better lower body power, speed, agility, and upper body power. In another study, the authors tried to predict selection in a talent identified junior volleyball squad with the physiological, anthropometric, and skill data [12]. They found that selected skill test results, not physiological and anthropometric data, could discriminate the successful and unsuccessful talent identified in junior volleyball players. However, the selected skill tests used only involved the subjective coach evaluation of passing and serving techniques. This evaluation probably will produce different results when it is applied by other coaches.

Whilst it has been generally acknowledged that volleyball players require a high level of physiological, physical, psychological as well as some specific anthropometric variables, the players also required a greater level of technical and tactical prowess for effective delivery of performance during the game [13]. Several skills such as serve, dig, set, and spike are shown to be the prerequisite for every volleyball player to master and execute efficiently. Although the mastery and possession of these skills could vary from one player to the other, nonetheless they are considered necessary at every level of expertise. It has been documented that the possessions of this playing ability could significantly distinguish novice, junior as well as elite players [14]. Moreover, as previously reported, the application of psychological strategies is crucial for the athletes as it promotes better coping skills during competitions [15, 16]; hence, the blend of physiological, fitness, psychological as well as technical and tactical skills is considered pivotal towards the successful delivery of performance in any sporting activity.

1.3 Recent Advances in Univariate and Machine Learning Application in Sports

The employment of machine learning in sporting activities has received due attention in recent years, primarily owing to the advent of Industry 4.0 [17]. The advancement of technology has pushed towards a more data-driven approach in a myriad of sporting applications. Machine learning (and even to a certain extend deep learning) has been explored in activity recognition, performance analysis, match outcome predictions, and even injury prediction amongst others [18]. This subsection shall provide a non-exhaustive brief review of recent works that use such an approach in sports.

The identification of different throws as well as the estimation of ball velocity in team handball was investigated by Tillaar et al. [19]. The authors utilized inertial measurement units (IMUs) that are strapped on the wrist of 17 handball players in acquiring the data of different throwing configurations. Conversely, a radar gun was used in order to capture the ball velocity. Two variations of the support vector machine (SVM) model, i.e. linear and polynomial, along with the random forest (RF) model as well as the gradient boosting machine (GBM) were used for both the classification and the regression problems. The feature selection technique was used to remove redundant features, whilst the tenfold cross-validation (CV) technique was employed to identify the optimal hyperparameters of the models developed. Owing to the size of the dataset, the leave-one-out (LOO) CV was used to estimate the classification accuracy of the models. It was demonstrated from the study that the polynomial-based SVM model was able to predict the ball velocity well with a mean absolute error (MAE) of 1.10 m/s, whilst the GBM model was able to classify the different throws reasonably well with a classification accuracy range between 79 and 87% on different classes.

Worsey et al. [20] investigated the employment of six different machine learning models, viz. logistic regression (LR), linear SVM, radial basis function (RBF)-SVM, RF, gradient boosting (XGB) as well as artificial neural networks (ANN) in the classification of running surface, specifically athletics track, hard sand, and soft sand. The data was collected through an IMU placed at the third thoracic vertebrae of seven able-bodied athletes that consists of four males and three females who were instructed to run at the different running surfaces at a distance of 400 m. Different statistical features were extracted from both time and frequency domain transformed signals. It is worth noting that the authors also employed principal component analysis (PCA) in order to reduce the dimensionality of the features. The authors investigated the effect of two CV techniques, i.e. LOO and the hold-out method, with a ratio of 75:25 for train and test data, respectively. It was shown from the study that the ANN model was able to distinguish well the classes based on the LOO CV, whilst RBF-SVM on the hold-out CV.

The paddle stroke identification of kayakers was investigated by Liu et al. [21]. Multiple IMUs were placed at different locations of the body of six female kayakers with an experience of at least three years. A number of statistical-based features were extracted from the signals captured via the IMUs. The dataset collected was split into a ratio of 80:20 before the extracted features are fed into different machine learning models, namely SVM, LR, decision tree (DT), k-nearest neighbour (kNN) and RF. It was reported that both the kNN and the SVM classifiers could attain an accuracy of up to 98.98% in recognizing the different paddle strokes.

The classification of skateboarding tricks by means of machine learning and deep learning models has also been explored in recent times [22–26]. Groh et al. are amongst the pioneers in investigating the use of machine learning in this particular sport. In [22], miPod IMU was fixed at the right front axis of the deck. Eleven predefined tricks were asked to be executed by 11 skateboarders. A total of 13 classes (11 tricks, 1 bail and 1 rest) were investigated with five classifiers, viz. NB, RF, linear SVM, RBF-SVM, and kNN. It was shown from the study that the RBF-SVM model could attain a classification accuracy (CA) of 89.1%. In an earlier study, Groh et al. [23], investigated the efficacy of four classifiers in classifying six skateboarding tricks executed by seven skateboarders. The same IMU configuration was used. Significant features were identified from the 54 features extracted prior to it was evaluated on the optimized classifiers (NB, partial decision tree (PART), SVM and kNN) via grid search. The best models were reported to be NB and SVM with a CA of 97.8 through LOO CV.

In a more recent study, Abdullah et al. [24] investigated the classification of five skateboarding tricks through the use of SVM, ANN, RF, kNN, NB and LR. The data was captured via a customized IMU placed at the front end of the skateboard, a similar configuration employed by [22]. The models were analysed via fivefold CV on time-domain features extracted from the IMU; it was shown that the LR and NB model could classify well the tricks with a CA of 95%. Conversely, Ibrahim et al. [25] extended the previous study by transforming the time domain signals to frequency domain via fast Fourier transform. Significant features were identified via the information gain feature selection technique from the transformed signals. It was shown that the NB model could attain a CA of 100% of the test dataset. The use of transfer learning architecture was investigated by Shapiee et al. [26] in the classification of five skateboarding tricks. The fully connected layer of the pretrained convolutional neural network (CNN); i.e., VGG-16 was replaced with LR and kNN models. The tricks were captured via a YI Action Camera, before the video was preprocessed to extract the images of the desired tricks. It was shown that the VGG-16-LR pipeline could attain a CA of 99.1% on the test dataset.

Chang et al. [27] investigated the use of a variation CNN architecture known as 'you only look once (YOLO) v3, particularly tiny YOLOv3 to capture the position of the puck of air hockey for the control of a real-time interactive system. The tiny YOLOv3 architecture has a reduced convolutional layer from 53 to 13 which has an average recognition speed of 30 fps. A total of 500 hockey images which consists of 250 clear images of the puck-moving slowly whilst the remaining 250 blurred images of the fast moving puck for training, whilst an additional 100 images were used for testing. It was shown that the model is able to recognize the puck at 94% accuracy.

It is worth mentioning that the identification of athlete performance class or talent identification has also been investigated through the use of machine learning. In contrast to the former literature surveyed, human-based performance indicators are used to identify the potential of the athletes [28–30]. For instance, the identification of high, medium, and low-performance sepak takraw players based on anthropometry indexes was investigated. The Louvain clustering technique was used to cluster the players based on the indexes [31]. It was shown from the study that a polynomial-based SVM model could provide an average CA of 96% based on the identified anthropometry indexes. ANN has also been used to distinguish between high and low-performance archers based on psychological attributes [16].

It has also been reported that ANN could provide a reasonable prediction (CA of up to 92.5%) on the win and lose of elite beach soccer team based on a number of technical and tactical performance indicators [32]. Therefore, it is evident from the brief literature survey that machine learning models have been demonstrated to be able to yield reasonable good prediction and deemed to be useful in the sporting domain.

1.4   Mann–Whiney U-Test Analysis

The Mann–Whitney U-test is a univariate-based mathematical analysis mainly used for means comparison. The U-test is a form of dependency tests analysis which presumes that the variables in the analysis can be separated into independent and dependent variables [33]. The rationale of applying the test is based on the assumptions that the variations in the average scores of the dependent variable(s) are largely attributed to the direct effect established by the independent variable(s). It is worth noting that the independent variable(s) is otherwise referred to as a factor since the factor normally divides the observed samples into two or more clusters. As opposed to the t-test and F-test, the Mann–Whitney U-test is considered as a nonparametric analysis which signifies that no prior assumption is established on the means of the distribution of the sample within the variables of interests in the population samples. In other words, the test is applied when the sample distributions are found to violate the normality assumptions [34].

The Mann–Whitney U-test has been successfully applied in different sports and has been shown to be effective in projecting differences between two levels of performance classes. For instance, in a recent study, the test has been applied to identify the technical as well as tactical performance indicators that could differentiate between the successful and unsuccessful team in elite beach soccer competitions [32]. In an earlier study, the Mann–Whitney U-test was also successfully employed to study the probability of sustaining sports injuries amongst British athletes partaking in wheelchair racing [35]. Similarly, the test has proved useful in ascertaining the differences in the learning styles of undergraduate athletic training students [36].

1.5   Features Extraction Analysis via Information Gain

Information gain (IG) is a common entropy-based function evaluation technique in machine learning. The IG method is most often used to derive information from one or more features in relation to a given categorical-dependent variable [37]. It is worth noting that in the present study IG is used to evaluate the functionality that can be used in providing information in order to measure the meaning of a specific variable for classification or discrimination tasks. The IG is used in the current investigation to extract information that demonstrates the importance of the functions, i.e. the performance parameters, in explaining the underlying associations with the performance of volleyball sport.

1.6 Cluster Analysis

1.6.1 Hierarchical Agglomerative Cluster Analysis (HACA)

Hierarchical agglomerative cluster analysis is commonly used as a tool of exploration as well as a non-exploratory method by which a cluster hierarchy for a single observation is established and a set of related observations form a distinct observation. [38]. It is important to note that in this algorithm the learning process is decided by the merges as well as the splits of the dataset, which are also implemented to isolate and illustrate identical findings in a dendrogram [15, 30, 39]. It should be remembered that HACA shows the number of clusters dependent on the vicinity of a given or predetermined clusters in the dendrogram. Distance of cosine was used in this analysis, and the clustering validation technique was conducted by class centroids [40].

1.6.2 Louvain Clustering

The Louvain clustering algorithm is also seen as the latest clustering algorithm that has the capability to group a given data or set of observations into a meaningful class. The algorithm is built to perform the said task following two distinctive steps; in the first step, it aims for a 'thin' group by maximizing the modularity in a traditional technique. In the second step, the algorithm puts together nodes of related communities and thereafter creates a distinctive community creating a new network of nodes of communities [41]. These processes can be performed iteratively before a modularity criterion is reached. This move also contributes to the hierarchical disintegration of the system and the formation of many partitions [42]. The divisions are typically the density of the communities' borders as opposed to the inter communities' edges.

1.7 Machine Learning Models

Machine learning is a subset of artificial intelligence, in which it is able to automatically improve and learn from past experiences without being explicitly programmed [17]. In essence, the algorithms are 'trained' to identify or recognize patterns and features from a reasonable size of data to allow it to yield better prediction based on fresh new data [43]. In this brief, both regression- and classification-based models (supervised machine learning models) are investigated. The readers are referred to our previous works [44, 45] on the definition of the different machine learning models as well as the performance indicators that is often employed in order to gauge the performance of a given model. The effect of hyperparameter tuning towards the prediction performance is also deliberated in some of the chapters.

1.8 Participants of the Study

The participants of this study consist of a total number of 24 teams that were involved in the UMT open volleyball tournament 2020. The teams that took part in the tournament comprised of UMT, PLUS, D' Volley, LPPVC, KVC, KK MAS, BTVC, JKR MARANG A, BnXBp, PDRM TRG, RISDA, DragonFLY, BaraVC, JKR MARANG B, BlockBuster, PKB, Venture VC, REDVELVET, Gheng Cheng, Pirates, Firefighter, Pulapol Dungun, USMKK, as well as FVC. It is important to highlight that each of the above-mentioned team is considered as an elite team with the players at least having an average of 6 years of volleyball playing experience. Moreover, a number of the players in some of the teams have already represented their states/country in both national and international competitions. The teams were assigned to a group of four each (Group A–F) in a typical congested fixture tournament schedule. The overall performances of the teams involved were notated and analysed throughout the two days competition period. A total number of 36 matches from the 24 teams were analysed. It is worth mentioning that before the beginning of the data collection, all the coaches, managers as well as the organizing committee were informed about the aim of the study and verbal consent was obtained via the Universiti Malaysia Terengganu, 2020 Volleyball Technical Committee (VBALL2020-UMT).

1.9 Performance Analysis

A total number of eight technical, as well as tactical performance parameters, were considered for evaluating the performance of all the individual player that plays for the team. The performance parameters that constitute Ace, Block, Set, Spike, Fault, Tap, Dig as well as Passing were used to analyse the players' performance in real time. It is worth noting that the selection of the aforesaid performance parameters was made based on their relevance to the game of volleyball as highlighted in some of the previous studies [46, 47]. A StatWatch application, a notational analysis framework built on android, was utilized as the device for assessing the performance of the players and the teams in accordance with the protocols previously reported by the past investigators [48]. Twelve experienced performance analysts were responsible for notating the performance of each player in a team such that each analyst was in charge of covering a particular player at a time. It is non-trivial to mention that the performance analysts were acquainted with the performance parameters chosen prior to the commencement of the full analysis. Using video from a separate match, reliability analysis was conducted. To maintain accuracy and test the observational errors on the selected performance parameters, the performance analysts were advised to notate the match separately, and their agreement was then compared. The Cohen's Kappa statistical test and Cronbach's alpha analysis were used to assess the analysts' consensus and consistency with respect to the performance parameters [49]. It is worth noting that a Kappa of 0.92 and a Cronbach's alpha of 0.96 were found, suggesting a high level of consensus and accuracy amongst the performance analysts in their overall analysis.

References

1. González-Ravé, J.M., Arija, A., Clemente-Suarez, V.: Seasonal changes in jump performance and body composition in women volleyball players. J. Strength Cond. Res. **25**, 1492–1501 (2011)
2. Fabian, T.: Volleygate: a history of scandal in the largest international sport federation. Sport Hist. Rev. **1**, 1–18 (2020)
3. Gabbett, T.J., Georgieff, B.: The development of a standardized skill assessment for junior volleyball players. Int. J. Sports Physiol. Perform. **1**, 95–107 (2006)
4. Polglaze, T., Dawson, B.: The physiological requirements of the positions in state league volleyball. Sport. Coach. **15**, 32 (1992)
5. Häkkinen, K.: Changes in physical fitness profile in female volleyball players during the competitive season. J. Sports Med. Phys. Fitness **33**, 223–232 (1993)
6. Abdullah, M.R., Musa, R.M., Kosni, N.A., Maliki, A.B.H.M., Haque, M.: Profiling and distinction of specific skills related performance and fitness level between senior and junior Malaysian youth soccer players. Int. J. Pharm. Res. **8**, 64–71 (2016)
7. Gipit, M.A., Charles, M.R.A., Musa, R.M., Kosni, N.A., Maliki, A.B.H.M.: The effectiveness of traditional games intervention programme in the improvement of form one school-age children's motor skills related performance components (2017)
8. Coutinho, P., Mesquita, I., Fonseca, A.M., Côte, J.: Expertise development in volleyball: the role of early sport activities and players' age and height. Kinesiology **47**, 215–225 (2015)
9. Spence, D.W., Disch, J.G., Fred, H.L., Coleman, A.E.: Descriptive profiles of highly skilled women volleyball players. Med. Sci. Sports Exerc. **12**, 299–302 (1980)
10. Milić, M., Grgantov, Z., Chamari, K., Ardigò, L.P., Bianco, A., Padulo, J.: Anthropometric and physical characteristics allow differentiation of young female volleyball players according to playing position and level of expertise. Biol. Sport. **34**, 19 (2017)
11. Paulo, A., Zaal, F.T.J.M., Fonseca, S., Araújo, D.: Predicting volleyball serve-reception. Front. Psychol. **7**, 1694 (2016)
12. Gabbett, T., Georgieff, B., Domrow, N.: The use of physiological, anthropometric, and skill data to predict selection in a talent-identified junior volleyball squad. J. Sports Sci. **25**, 1337–1344 (2007)
13. Eloi, S., Langlois, V., Jarrett, K.: The role of the Libero in volleyball as a paradoxical influence on the game: logical debate and the proposal for a rule change. Sport J. (2015)

14. Thissen-Milder, M., Mayhew, J.L.: Selection and classification of high school volleyball players from performance tests. J. Sports Med. Phys. Fitness **31**, 380–384 (1991)
15. Maliki, A.B.H.M., Abdullah, M.R., Juahir, H., Abdullah, F., Abdullah, N.A.S., Musa, R.M., Mat-Rasid, S.M., Adnan, A., Kosni, N.A., Muhamad, W.S.A.W., Nasir, N.A.M.: A multilateral modelling of youth soccer performance index (YSPI). IOP Conf. Ser. Mater. Sci. Eng. **342**, (2018). https://doi.org/10.1088/1757-899X/342/1/012057
16. Musa, R.M., Taha, Z., Majeed, A.P.P.A., Abdullah, M.R.: Psychological variables in ascertaining potential archers. In: Machine Learning in Sports, pp. 21–27. Springer (2019)
17. Richter, C., O'Reilly, M., Delahunt, E.: Machine learning in sports science: challenges and opportunities. Sport. Biomech. 1–7 (2021). https://doi.org/10.1080/14763141.2021.1910334
18. Van Eetvelde, H., Mendonça, L.D., Ley, C., Seil, R., Tischer, T.: Machine learning methods in sport injury prediction and prevention: a systematic review. J. Exp. Ortop. **8**, 27 (2021). https://doi.org/10.1186/s40634-021-00346-x
19. Van Den Tillaar, R., Bhandurge, S., Stewart, T.: Can Machine Learning with IMUs Be Used to Detect Different Throws and Estimate Ball Velocity in Team Handball? (2021). https://doi.org/10.3390/s21072288
20. Worsey, M.T.O., Espinosa, H.G., Shepherd, J.B., Thiel, D. V.: One size doesn't fit all: supervised machine learning classification in athlete-monitoring. IEEE Sens. Lett. **5** (2021). https://doi.org/10.1109/LSENS.2021.3060376
21. Liu, L., Wang, H.-H., Qiu, S., Zhang, Y.-C., Hao, Z.-D., Zhang, S., Hao, Y.-C., Paddle, Z.-D.: Paddle stroke analysis for kayakers using wearable technologies (2021). https://doi.org/10.3390/s21030914
22. Groh, B.H., Fleckenstein, M., Kautz, T., Eskofier, B.M.: Classification and visualization of skateboard tricks using wearable sensors. Pervasive Mob. Comput. **40**, 42–55 (2017)
23. Groh, B.H., Kautz, T., Schuldhaus, D.: IMU-based trick classification in skateboarding. KDD Work. Large-Scale Sport. Anal. (2015)
24. Abdullah, M.A., Ibrahim, M.A.R., Shapiee, M.N.A. Bin, Mohd Razman, M.A., Musa, R.M., Abdul Majeed, A.P.P.: The classification of skateboarding trick manoeuvres through the integration of IMU and machine learning. Presented at the (2020). https://doi.org/10.1007/978-981-13-9539-0_7
25. Ibrahim, M.A.R., Shapiee, M.N.A., Abdullah, M.A., Razman, M.A.M., Musa, R.M., Majeed, A.P.P.A.: The classification of skateboarding trick manoeuvres: a frequency-domain evaluation. In: Embracing Industry 4.0, pp. 183–194. Springer (2020)
26. Shapiee, M.N.A., Ibrahim, M.A.R., Razman, M.A.M., Abdullah, M.A., Musa, R.M., Majeed, A.P.P.A.: The classification of skateboarding tricks by means of the integration of transfer learning and machine learning models. In: Embracing Industry 4.0, pp. 219–226. Springer (2020)
27. Chang, C.-L., Chen, S.-T., Chang, C.-Y., Jhou, Y.-C.: Application of machine learning in air hockey interactive control system. https://doi.org/10.3390/s20247233
28. Taha, Z., Musa, R.M., P.P. Abdul Majeed, A., Alim, M.M., Abdullah, M.R.: The identification of high potential archers based on fitness and motor ability variables: a support vector machine approach. Hum. Mov. Sci. **57**, 184–193 (2018). https://doi.org/10.1016/j.humov.2017.12.008
29. Musa, R.M., Taha, Z., Majeed, A.P.P.A., Abdullah, M.R.: Machine Learning in Sports: Identifying Potential Archers. Springer (2019)
30. Musa, R.M., Abdullah, M.R., Maliki, A.B.H.M., Kosni, N.A., Mat-Rasid, S.M., Adnan, A., Juahir, H.: Supervised pattern recognition of archers' relative psychological coping skills as a component for a better archery performance. J. Fundam. Appl. Sci. **10**, 467–484 (2018)
31. Musa, R.M., Majeed, A.P.P.A., Kosni, N.A., Abdullah, M.R.: Identifying talent in sepak takraw via anthropometry indexes. In: Machine Learning in Team Sports, pp. 29–39. Springer (2020)
32. Musa, R.M., Majeed, A.P.P.A., Kosni, N.A., Abdullah, M.R.: Technical and tactical performance indicators determining successful and unsuccessful team in elite beach soccer. In: Machine Learning in Team Sports, pp. 21–28. Springer (2020)

33. Muazu Musa, R., PP Abdul Majeed, A., Abdullah, M.R., Ab. Nasir, A.F., Arif Hassan, M.H., Mohd Razman, M.A.: Technical and tactical performance indicators discriminating winning and losing team in elite Asian beach soccer tournament. PLoS One. **14**, e0219138 (2019)
34. MacFarland, T.W., Yates, J.M.: Mann–whitney u test. In: Introduction to nonparametric statistics for the biological sciences using R, pp. 103–132. Springer (2016)
35. Taylor, D., Williams, T.: Sports injuries in athletes with disabilities: wheelchair racing. Spinal Cord. **33**, 296–299 (1995)
36. Brower, K.A., Stemmans, C.L., Ingersoll, C.D., Langley, D.J.: An investigation of undergraduate athletic training students' learning styles and program admission success. J. Athl. Train. **36**, 130 (2001)
37. Musa, R.M., Abdul Majeed, A.P.P., Musa, A., Abdullah, M.R., Kosni, N.A., Razman, M.A. M.: An information gain and hierarchical agglomerative clustering analysis in identifying key performance parameters in elite beach soccer. Presented at the (2021). https://doi.org/10.1007/978-981-15-7309-5_26
38. Maimon, O., Rokach, L.: Data mining and knowledge discovery handbook. (2005). https://doi.org/10.1007/b107408
39. Razali, M.R., Alias, N., Maliki, A., Musa, R.M., Kosni, L.A., Juahir, H.: Unsupervised pattern recognition of physical fitness related performance parameters among Terengganu youth female field hockey players. Int. J. Adv. Sci. Eng. Inf. Technol. **7**, 100–105 (2017)
40. Muazu Musa, R., Abdul Majeed, A.P.P., Taha, Z., Abdullah, M.R., Husin Musawi Maliki, A. B., Azura Kosni, N.: The application of Artificial Neural Network and k-Nearest Neighbour classification models in the scouting of high-performance archers from a selected fitness and motor skill performance parameters. Sci. Sport. (2019). https://doi.org/10.1016/j.scispo.2019.02.006
41. Wu, C., Gudivada, R.C., Aronow, B.J., Jegga, A.G.: Computational drug repositioning through heterogeneous network clustering. BMC Syst. Biol. **7**, S6 (2013). https://doi.org/10.1186/1752-0509-7-S5-S6
42. Blondel, V.D., Guillaume, J., Lambiotte, R., Lefebvre, E.: Fast unfolding of community hierarchies in large networks. J. Stat. Mech. theory Exp. 10008 (2008). https://doi.org/10.1088/1742-5468/2008/10/P10008
43. Taha, Z., Razman, M.A.M., Adnan, F.A., Abdul Ghani, A.S., Abdul Majeed, A.P.P., Musa, R.M., Sallehudin, M.F., Mukai, Y.: The identification of hunger behaviour of lates calcarifer through the integration of image processing technique and support vector machine. In: IOP Conference Series: Materials Science and Engineering (2018). https://doi.org/10.1088/1757-899X/319/1/012028
44. Muazu Musa, R.P.P., Abdul Majeed, A., Kosni, N.A., Abdullah, M.R.: Machine Learning in Team Sports. Springer Singapore, Singapore (2020). https://doi.org/10.1007/978-981-15-3219-1
45. Muazu Musa, R., Taha, Z.P.P., Abdul Majeed, A., Abdullah, M.R.: Machine learning in sports. (2019). https://doi.org/10.1007/978-981-13-2592-2
46. Education, C.: Coaching volleyball technical and tactical skills. Hum. Kinetics (2011)
47. Giddens, S., Giddens, O.: Volleyball: Rules, Tips, Strategy, and Safety. The Rosen Publishing Group (2005)
48. Abdullah, M.R., Musa, R.M., Maliki, A.B.H.M., Kosni, N.A., Suppiah, P.K.: Development of tablet application based notational analysis system and the establishment of its reliability in soccer. J. Phys. Educ. Sport. **16**, 951–956 (2016). https://doi.org/10.7752/jpes.2016.03150
49. McGuigan, K., Hughes, M., Martin, D.: Performance indicators in club level Gaelic football. Int. J. Perform. Anal. Sport. **18**, 780–795 (2018). https://doi.org/10.1080/24748668.2018.1517291

Chapter 2
The Effect of Competition Strategies in Influencing Volleyball Performance

Abstract In this chapter, we examine the importance of key competitive psychological strategic elements that could be used to identify high-performance players in elite indoor men volleyball tournaments. It is demonstrated that a set of competitional psychological strategic elements, namely self-talk, activation, imagery, emotion control, automaticity, relaxation, goal setting, as well as negative thinking, are essential in determining performance and consequently serve as the indicators that significantly favour high performance in the elite volleyball game. In addition, it was demonstrated that a high classification accuracy in discerning the class, i.e. the performance of the players, could be attained via the k-nearest neighbours classifier.

Keywords Psychological coping skill · Indoor volleyball · High-performance players · k-nearest neighbours classifier

2.1 Overview

During the sporting competition, athletes are expected to deliver the best performance irrespective of the physiological and psychological factors inherent in the sport. The ability of an athlete to perform under any circumstances defines the worth of the athlete in the team. Moreover, it is worth noting that sporting competition, especially at the elite level, is undertaken in an environment that is characterized by a high level of stress and arousal due to the desire attach to winning in any competition [1]. The stress involved is widely used to describe an uncomfortable mental situation or disorder marked by subjective feelings of tension, fear, and concern [2]. It is generally referred to as precompetition stress or fear in athletics. Furthermore, studies have shown that fear has a negative impact on these sports outcomes [3]. Therefore, the usage of certain psychological strategies to curtail the aforesaid stressors as well as improve the performance in sports has significantly evolved in the contemporary sporting domain [4]. The application of such psychological strategies is crucial for the athletes as it promotes better coping skills during competitions [5, 6].

The nature of the volleyball sport as an intermittent as well as a non-inversion game that is characterized by short bouts of great intensity movement interwoven by a considerable low level of activity has underscored the need of the players to possess a high level of physical fitness [7, 8]. Nonetheless, it is important to note that the characteristics of the game and its facilities such as the physical attributes of the players, the size of the net, and the court, as well as the fixture of the match, demands a high level of psychological skill. During the competition, the afore-mentioned factors present room for a little mistake that could be detrimental to a team and potentially alter the outcome of the match [9]. However, despite the apparent importance of the possible association of the mental fitness aspect towards the successful delivery of performance in this sport, little is known about the contribution of such psychological strategies in influencing volleyball performance at a highly competitive level. Thus, the present study is carried out to ascertain the influence of competitional psychological strategies in the performance of elite volleyball players.

Competitional strategies assessment: The performance strategies–competition scale (TOPS-SC) originally developed and validated by the preceding researchers was utilized to assess the psychological elements of the players in the study [10]. The TOPS-SC assessed eight psychological coping strategies during competition. These coping strategies constituted self-talk, activation, imagery, emotion control, automaticity, relaxation, goal setting as well as negative thinking. The players completed this inventory, and their performances throughout the tournament were analysed as described in the previous chapter. It is worth highlighting that the players who are unable to play at least 70% of the matches in the team were not considered for the final analysis in this study.

2.2   Clustering

The Louvain clustering technique was employed for the clustering analysis in this study. The performance of the players, as well as the scores of the players in the competitional strategic elements, viz. self-talk, activation, imagery, emotion control, automaticity, relaxation, goal setting, as well as negative thinking, was used as the input variables for the clustering analysis. It is important to mention that the clustering analysis is essentially applied to ascertain the performance group of the players in order to give room for the assignation of class membership based on all the aforementioned variables [11, 12].

2.3   Classification

In the present investigation, the k-nearest neighbours classifier was investigated towards its efficacy in classifying the performance of the players. The dataset was split into a 70:30 ratio for training and testing, respectively [13]. The number of neighbours was varied between one and 25, whilst the distance metrics employed is Minkowski. The identification of the best hyperparameter, i.e. the number of neighbours, was carried out via a fivefold cross-validation technique within the training dataset [14]. The performance of the model is demonstrated through the classification accuracy as well as the confusion matrix. The evaluation of the model is carried out on Spyder IDE (Python 3.6) through scikit-learn libraries.

2.4   Results and Discussion

Figure 2.1 depicts the classes assigned by the Louvain clustering analysis with respect to the assessed variables in the study. It could be observed from the figure that two classes are discovered based on the peculiarity of the players in the performance of the variables evaluated. The figure further shows a simple allocation of group memberships that aids in the assignment of group tag to each class, namely highly competitive strategic (HCS) players and low competitive strategic (LCS) players, respectively.

Table 2.1 shows the inferential statistics on the performance differences between the groups based on the psychological elements evaluated as projected by the Mann–Whitney analysis. The test is carried out to ascertain the competitional strategic element that significantly differentiates the two groups [15]. It could be seen from the table that the HCS group possesses significantly higher levels of the competitional strategic elements as opposed to the LCS group $p < 0.001$.

It could be observed from Fig. 2.2 that the optimum value of k that is able to yield a reasonably accurate classification of the performance of the players is 6 with an average classification accuracy (CA) of 0.954 (± 0.123) via the cross-validation evaluation. The model is then evaluated in the test dataset, and it was found that a CA of 96.4% was achieved. Figure 2.3 illustrates the confusion matrix on the test dataset.

The findings from the present investigation revealed that successful performance in the sport of volleyball is reliant upon several mental fitness elements. In other words, the needs for several competitional strategic elements towards the successful performance of a volleyball game could not be overemphasized as evident from Fig. 2.1 and Table 2.1. This could be further deduced as the demands for various forms of sports skills and pieces of training have greatly increased and the distinctions of the gap between the players on physical performance as well as records have seamlessly condensed; thus, psychological abilities have become increasingly important in such a way that several coaches and team managers stress the significance of psychological abilities towards the attainment of athletic excellence [1, 3, 4, 9, 16, 17].

The study findings further revealed that at the elite level of volleyball sport, the most successful players are distinguished by the ability to harness all the

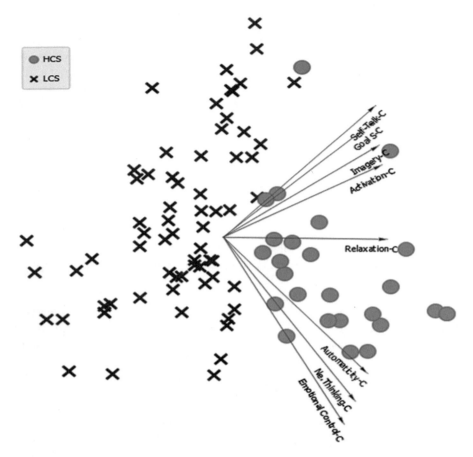

Fig. 2.1 Class affiliation allocated by the Louvain clustering algorithm

Table 2.1 Inferential differences between the HCS and LCS on the measured competition strategic elements

Competition strategy elements	LCS		HCS		Sig (MWU-test)
	Mean	Std.D	Mean	Std.D	
Goal setting	3.765	0.603	4.538	0.422	0.001*
Automaticity	3.045	0.576	4.192	0.571	0.001*
Emotional control	3.011	0.588	4.144	0.875	0.001*
Imagery	3.649	0.602	4.365	0.460	0.001*
Activation	3.627	0.580	4.365	0.426	0.001*
Self-talk	3.724	0.640	4.558	0.396	0.001*
Relaxation	3.437	0.479	4.394	0.459	0.001*
Negative thinking	2.896	0.463	4.058	0.558	0.001*

*$P < 0.001$

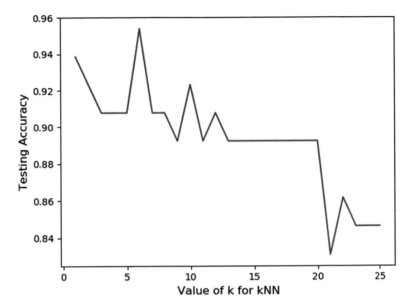

Fig. 2.2 The identification of the optimum *k*

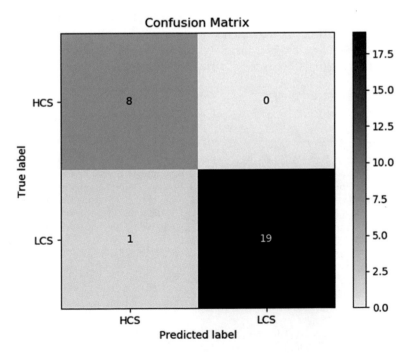

Fig. 2.3 Confusion matrix on the test dataset

competitional coping strategies elements. The findings of this study are unanimous with the previous investigations which demonstrated that the successful elite volleyball players possessed a better mental fitness as opposed to the unsuccessful ones [17]. The authors further inferred that the elite players are attributed with a high level of self-confidence, better concentration as well mental toughness which shield them from being affected by emotions and hitherto demonstrate more successful delivery of performance.

2.5   Summary

The present investigation has established the key competitional psychological strategic elements that could be used to identify high-performance players in elite indoor volleyball tournaments. A set of competitional psychological strategic elements, namely self-talk, activation, imagery, emotion control, automaticity, relaxation, goal setting, as well as negative thinking, are demonstrated to be essential in determining performance and consequently serve as the indicators that significantly favour high performance in the men elite volleyball game. In addition, it was demonstrated that a high classification accuracy in discerning the class, i.e. the performance of the players, could be attained via the k-nearest neighbours classifier. Moreover, it has been shown that high-performance players at the elite level in this sport are not free from negative thinking, indicating that the players are characterized by worry over the outcome of a match. This should be considered when applying intervention strategy to help them cope better.

References

1. Taha, Z., Musa, R.M., Abdullah, M.R., Maliki, A., Kosni, N.A., Mat-Rasid, S.M., Adnan, A., Juahir, H.: Supervised pattern recognition of archers' relative psychological coping skills as a component for a better archery performance. J. Fundam. Appl. Sci. **10**, 467–484 (2018)
2. Musa, R.M., Abdullah, M.R., Juahir, H., Eswaramoorthi, V., Alias, N., Hashim, M.R., Alnamat, A.S.F.: An exploratory study of personality traits and psychological coping skills on archery performance. Indian J. Public Heal. Res. Dev. **10**, 630–635 (2019). https://doi.org/10. 5958/0976-5506.2019.00572.2
3. John, S., Verma, S.K., Khanna, G.L.: The effect of mindfulness meditation on HPA-Axis in pre-competition stress in sports performance of elite shooters. Natl. J. Integr. Res. Med. **2**, 15–21 (2011)
4. Abdullah, M.R., Musa, R.M., Maliki, A.B.H.M.B., Kosni, N.A., Suppiah, P.K.: Role of psychological factors on the performance of elite soccer players. J. Phys. Educ. Sport. **16**, 170 (2016)
5. Maliki, A.B.H.M., Abdullah, M.R., Juahir, H., Abdullah, F., Abdullah, N.A.S., Musa, R.M., Mat-Rasid, S.M., Adnan, A., Kosni, N.A., Muhamad, W.S.A.W., Nasir, N.A.M.: A multilateral modelling of youth soccer performance index (YSPI). IOP Conf. Ser. Mater. Sci. Eng. **342**, (2018). https://doi.org/10.1088/1757-899X/342/1/012057

6. Musa, R.M., Taha, Z., Majeed, A.P.P.A., Abdullah, M.R.: Psychological variables in ascertaining potential archers. In: Machine Learning in Sports, pp. 21–27. Springer (2019)
7. Gabbett, T.J., Georgieff, B.: The development of a standardized skill assessment for junior volleyball players. Int. J. Sports Physiol. Perform. **1**, 95–107 (2006)
8. Polglaze, T., Dawson, B.: The physiological requirements of the positions in state league volleyball. Sport. Coach. **15**, 32 (1992)
9. Shea, C.H., Kohl, R.M.: Composition of practice: Influence on the retention of motor skills. Res. Q. Exerc. Sport **62**, 187–195 (1991)
10. Thomas, P.R., Murphy, S.M., Hardy, L.E.W.: Test of performance strategies: Development and preliminary validation of a comprehensive measure of athletes' psychological skills. J. Sports Sci. **17**, 697–711 (1999)
11. Musa, R.M., Majeed, A.P.P.A., Kosni, N.A., Abdullah, M.R.: An overview of beach soccer, sepak takraw and the application of machine learning in team sports. Mach. Learn. Team Sport. 1–12 (2020)
12. Razali, M.R., Alias, N., Maliki, A., Musa, R.M., Kosni, L.A., Juahir, H.: Unsupervised pattern recognition of physical fitness related performance parameters among Terengganu youth female field hockey players. Int. J. Adv. Sci. Eng. Inf. Technol. **7**, 100–105 (2017)
13. Taha, Z., Razman, M.A.M., Adnan, F.A., Abdul Ghani, A.S., Abdul Majeed, A.P.P., Musa, R.M., Sallehudin, M.F., Mukai, Y.: The identification of hunger behaviour of lates calcarifer through the integration of image processing technique and support vector machine. In: IOP Conference Series: Materials Science and Engineering (2018). https://doi.org/10.1088/1757-899X/319/1/012028
14. Musa, R.M., Majeed, A.P.P.A., Taha, Z., Abdullah, M.R., Maliki, A.B.H.M., Kosni, N.A.: The application of artificial neural network and k-nearest neighbour classification models in the scouting of high-performance archers from a selected fitness and motor skill performance parameters. Sci. Sports **34**, e241–e249 (2019)
15. Muazu Musa, R.P.P., Abdul Majeed, A., Abdullah, M.R., Ab. Nasir, A.F., Arif Hassan, M.H., Mohd Razman, M.A.: Technical and tactical performance indicators discriminating winning and losing team in elite Asian beach soccer tournament. PLoS One. **14**, e0219138 (2019). https://doi.org/10.1371/journal.pone.0219138
16. Abdullah, M.R., Maliki, A.B.H.M.: Reliability of test of performance strategies-competition scale (TOPS-CS) among… Man India. **96**, 5199–5207 (2016)
17. Mohammadzadeh, H., Sami, S.: Psychological skills of elite and non-elite volleyball players. Ann. Appl. Sport Sci. **2**, 31–36 (2014)

Chapter 3
Identification of Psychological Training Strategies Essential for Volleyball Performance

Abstract This chapter highlights the contribution of certain training psychological strategic elements in the prediction of players performance during the indoor volleyball tournament. It has been demonstrated from the study findings that training strategic variables that included attentional control, imagery, relaxation, activation, goal setting, emotional control, automaticity as well as self-talk are found to be non-trivial in predicting the performance of the elite volleyball players. It has also been demonstrated from the current finding that linear-based SVR is able to yield an accurate prediction of the performance of the players. It is then inferred that the ability of an athlete to develop a personal strategy for dealing with adversity could to a larger extend assist the athlete in delivering good performance during competition in this sport.

Keywords Training strategy · Coping with adversity · Indoor volleyball · Regression analysis

3.1 Overview

Optimal performance in the sporting domain largely depends on physical ability, skills as well as psychological factors. Physical ability refers to those key attributes that are pertinent to the type of sport an athlete is involved in. For instance, the fitness level, the right physique to make an impact in a specific role all constitutes physical ability, whilst the skill involves the possessions of several technical as well as tactical position-related resources to play at the highest level coupled with the ability to make strategic on-demand thinking, read the game, and quickly make decisions that could lead to success [1]. Research has demonstrated that physical fitness, as well as technical and tactical ability, is pivotal to the successful delivery of performance in sporting activity [2–6]. However, recent evidence projected that the possession of the physical ability, technical as well as tactical skills alone could not ensure success without the blend of some psychological strategies [7]. It is important to note that the ability of a player to cope with the daily pressure,

© The Author(s), under exclusive license to Springer Nature Singapore Pte Ltd. 2021   21
R. Muazu Musa et al., *Machine Learning in Elite Volleyball*, SpringerBriefs
in Applied Sciences and Technology, https://doi.org/10.1007/978-981-16-3192-4_3

concentrate in-game and training situations, the desire to keeping working and positive interactions with both the coaches and the teammates not only ensure success but also facilitate career development path for the athletes [8].

According to study results, competitors experience more performance-related stressors prior to competing [9]. These findings emphasize the importance of considering all the pressures placed on athletes when planning and executing training programmes. This pointed to the needs for interventions to relieve the competition burden since precomputation anxiety is a common ailment that occurs amongst athletes of all levels and in every sport [10–12]. Nonetheless, it has been reported that psychological strategies are wide-ranging and cut across several areas in the sport dimension. For instance, some are developed for the initiation, control, and management of motivation [13], whilst others are primarily built to strengthen concentration, self-confidence as well as regulations of emotions [14]. Like many other non-inversion sports, the sport of volleyball is characterized by a high level of psychological demands as a result of the fast-paced nature of the sport. Thus, it is non-trivial to identify the essential psychological traits that could guarantee success in the sport, especially at the elite level where the stakes are high. Hence, the purpose of the current study is to identify the essential psychological training strategies that could predict the performance of the players in this sport.

Evaluation of the training strategies: The previously developed and tested tool for training strategies—scale (TOPS-PS)—was used to measure the psychological aspects of the players in the sample [15, 16]. The TOPS-PS evaluated eight psychological coping mechanisms during training, namely, attentional control, imagery, relaxation, activation, goal setting, emotional control, automaticity as well as self-talk. The inventory was completed by the players, and their results during the tournament were evaluated as stated in the previous chapter. It is worth remembering that the players who are unable to play at least 70% of the time were not considered for the final analysis in this investigation.

3.2   Feature Selection

A feature selection technique was employed to extract the vital psychological training strategies relevant to the sport of volleyball. It is worth highlighting that such a technique is non-trivial in order to reduce overfitting, improve accuracy as well as reduce the training time [17]. In this study, the psychological training strategies scores of the players (attentional control, imagery, relaxation, activation, goal setting, emotional control, automaticity as well as self-talk) were evaluated on different feature selection techniques, namely information gain (IG), gain ratio, ReliefF as well as chi-square analysis.

3.3   Machine Learning-Based Regression Model

In the present study, a variety of support vector machine regression [18] models, known as support vector regressors (SVRs), is evaluated towards its ability to predict the performance of the players. A total of 93 data were split to a ratio of 70:30 for training and testing, respectively [19, 20]. The different types of SVR kernels are investigated, namely linear, cubic and radial basis function (RBF). The models were developed based on the significant features identified via the feature selection method described previously. The performance of the models is measured via the coefficient of determination, R^2, and the mean absolute error, (MAE). Spyder IDE (Python 3.6) was used to develop and analyse the efficacy of the models.

3.4   Results and Discussion

Table 3.1 shows the feature extractions technique employed to identify the essential training strategic elements influencing the performance of volleyball. It could be seen from the table that the most significant features identified from the IG features selection technique are attentional control, imagery, relaxation, activation as well as goal settings.

Based on significant features identified via the IG, the linear, cubic and RBF SVRs are able to yield a test R^2 of 0.9999, 0.8262, and 0.6339 with a test MAE of 0.0616, 6.102, and 4.4697, respectively. Therefore, it could be concluded that the linear-based SVR is able to yield an accurate prediction of the performance of the players. Figure 3.1 depicts the prediction values of the developed model again the actual data.

Psychological training strategic elements have been reported to be pivotal in improving athletic performance in various sports. For instance, some psychological elements such as stress and worry were found to be detrimental to athletic performance [21–23]. On the other hand, emotional control has been indicated to be an integral aspect in ensuring victory during sporting context [24]. The authors further

Table 3.1 Feature selection

Training strategy elements	Infor. gain
Attentional control	**0.103**
Imagery	**0.063**
Relaxation	**0.054**
Activation	**0.053**
Goal setting	**0.050**
Emotional control	0.049
Automaticity	0.036
Self-talk	0.036

A threshold value of above or equal to 0.05 is considered to be significant

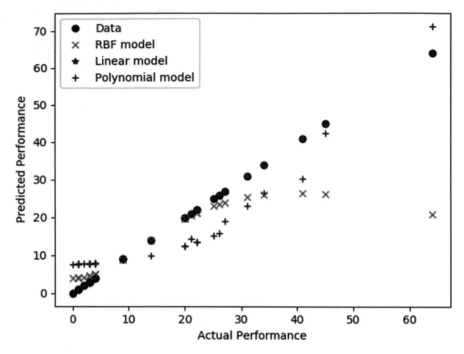

Fig. 3.1 Prediction values of the developed model again the actual data

inferred that athletes' emotional state such as happiness, joy, excitements as well as sadness emanating from a crushing defeat may affect athletic performance and the outcome of a competition. Hence, the ability of an athlete to control his or her emotion could result in positive performance outcome both during training and during competition. It has been established from the current investigation that the training strategic elements that constitute attentional control, imagery, relaxation, activation, goal setting, emotional control, automaticity as well as self-talk are crucial in predicting the performance of elite volleyball players. These psychological elements could be termed as the individual personal–psychological strategy that describes the quality of oneself in dealing with varying degree of adversity during a sporting performance.

3.5 Summary

The present investigation has established the psychological training strategies that are essential in predicting the performance of the elite indoor volleyball players during the tournament. These sets of psychological training elements that included attentional control, imagery, relaxation, activation, goal setting, emotional control, automaticity as well as self-talk are found to be non-trivial in predicting the performance of the elite volleyball players. It has also been demonstrated from the current finding that linear-based SVR is able to yield an accurate prediction of the performance of the players. Moreover, the ability of an athlete to develop a personal strategy for dealing with adversity could to a larger extend assist the athlete in delivering good performance during competition in this sport. This point should be considered when developing a personal–psychological strategic plan for the players in elite volleyball sport.

References

1. Durand-Bush, N., Salmela, J.H.: The development and maintenance of expert athletic performance: perceptions of world and Olympic champions. J. Appl. Sport Psychol. **14**, 154–171 (2002)
2. Razali, M.R., Alias, N., Maliki, A., Musa, R.M., Kosni, L.A., Juahir, H.: Unsupervised pattern recognition of physical fitness related performance parameters among Terengganu youth female field hockey players. Int. J. Adv. Sci. Eng. Inf. Technol. **7**, 100–105 (2017)
3. Abdullah, M.R., Musa, R.M., Kosni, N.A., Maliki, A.B.H.M., Haque, M.: Profiling and distinction of specific skills related performance and fitness level between senior and junior Malaysian youth soccer players. Int. J. Pharm. Res. (2016)
4. Taha, Z., Haque, M., Musa, R.M., Abdullah, M.R., Maliki, A., Alias, N., Kosni, N.A.: Intelligent prediction of suitable physical characteristics toward archery performance using multivariate techniques. J. Glob. Pharm. Technol. **9**, 44–52 (2009)
5. Gipit, M.A., Charles, M.R.A., Musa, R.M., Kosni, N.A., Maliki, A.B.H.M.: The effectiveness of traditional games intervention programme in the improvement of form one school-age children's motor skills related performance components (2017)
6. Muazu Musa, R.P.P., Abdul Majeed, A., Abdullah, M.R., Ab. Nasir, A.F., Arif Hassan, M.H., Mohd Razman, M.A.: Technical and tactical performance indicators discriminating winning and losing team in elite Asian beach soccer tournament. PLoS One **14**, e0219138 (2019). https://doi.org/10.1371/journal.pone.0219138
7. Abdullah, M.R., Musa, R.M., Maliki, A.B.H.M.B., Kosni, N.A., Suppiah, P.K.: Role of psychological factors on the performance of elite soccer players. J. Phys. Educ. Sport. **16**, 170 (2016)
8. Gabbett, T., Georgieff, B., Domrow, N.: The use of physiological, anthropometric, and skill data to predict selection in a talent-identified junior volleyball squad. J. Sports Sci. **25**, 1337–1344 (2007)
9. John, S., Verma, S.K., Khanna, G.L.: The effect of mindfulness meditation on HPA-Axis in pre-competition stress in sports performance of elite shooters. Natl. J. Integr. Res. Med. **2**, 15–21 (2011)
10. Chamberlain, S.T., Hale, B.D.: Competitive state anxiety and self-confidence: intensity and direction as relative predictors of performance on a golf putting task. Anxiety Stress Coping **20**, 197–207 (2007)
11. Kais, K., Raudsepp, L.: Intensity and direction of competitive state anxiety, self-confidence and athletic performance. Kinesiology **37**, 13–20 (2005)
12. Swain, A., Jones, G.: Explaining performance variance: the relative contribution of intensity and direction dimensions of competitive state anxiety. Anxiety Stress Coping **9**, 1–18 (1996)
13. Reiss, S., Wiltz, J., Sherman, M.: Trait motivational correlates of athleticism. Pers. Individ. Dif. **30**, 1139–1145 (2001)

14. Bois, J.E., Sarrazin, P.G., Southon, J., Boiché, J.C.S.: Psychological characteristics and their relation to performance in professional golfers. Sport Psychol. **23**, 252–270 (2009)
15. Abdullah, M.R., Maliki, A.B.H.M.: Reliability of test of performance strategies-competition scale (TOPS-CS) among… Man India. **96**, 5199–5207 (2016)
16. Thomas, P.R., Murphy, S.M., Hardy, L.E.W.: Test of performance strategies: development and preliminary validation of a comprehensive measure of athletes' psychological skills. J. Sports Sci. **17**, 697–711 (1999)
17. Musa, R.M., Abdul Majeed, A.P.P., Musa, A., Abdullah, M.R., Kosni, N.A., Razman, M.A. M.: An Information Gain and hierarchical agglomerative clustering analysis in identifying key performance parameters in elite beach soccer. Presented at the (2021). https://doi.org/10.1007/978-981-15-7309-5_26
18. Radzuan, N.Q., Hassan, M.H.A., Abdul Majeed, A.P.P., Musa, R.M., Mohd Razman, M.A., Abu Kassim, K.A.: Predicting serious injuries due to road traffic accidents in Malaysia by means of artificial neural network BT—intelligent manufacturing and mechatronics. Presented at the (2020)
19. Taha, Z., Razman, M.A.M., Adnan, F.A., Abdul Ghani, A.S., Abdul Majeed, A.P.P., Musa, R.M., Sallehudin, M.F., Mukai, Y.: The identification of hunger behaviour of lates calcarifer through the integration of image processing technique and support vector machine. In: IOP Conference Series: Materials Science and Engineering (2018). https://doi.org/10.1088/1757-899X/319/1/012028
20. Musa, R.M., Majeed, A.P.P.A., Taha, Z., Abdullah, M.R., Maliki, A.B.H.M., Kosni, N.A.: The application of artificial neural network and k-nearest neighbour classification models in the scouting of high-performance archers from a selected fitness and motor skill performance parameters. Sci. Sports **34**, e241–e249 (2019)
21. Lazarus, R.S.: How emotions influence performance in competitive sports. Sport Psychol. **14**, 229–252 (2000)
22. Musa, R.M., Abdullah, M.R., Juahir, H., Eswaramoorthi, V., Alias, N., Hashim, M.R., Alnamat, A.S.F.: An exploratory study of personality traits and psychological coping skills on archery performance. Indian J. Public Heal. Res. Dev. (2019). https://doi.org/10.5958/0976-5506.2019.00572.2
23. Musa, R.M., Taha, Z., Majeed, A.P.P.A., Abdullah, M.R.: Psychological variables in ascertaining potential archers. In: Machine Learning in Sports, pp. 21–27. Springer (2019)
24. Jones, M.V.: Controlling emotions in sport. Sport Psychol. **17**, 471–486 (2003)

Chapter 4
The Strategic Competitional Elements Contributing to Volleyball Performance

Abstract This chapter explored the importance of certain psychological strategic elements in the prediction of players performance during the indoor volleyball tournament. It has been demonstrated from the study findings that competitional strategic variables that constituted imagery, self-talk, activation, automaticity as well as emotional control are shown to be pivotal in determining the performance of elite volleyball players during competition. It is then postulated that the ability of an athlete to develop a personal strategy for coping with obstacles will help the athlete produce better results during competition in this sport to a greater extent.

Keywords Competitional strategy · Indoor volleyball · Coping skill · Regression model

4.1 Overview

Athletes are encouraged to give their best results during athletic competitions, regardless of the physiological and psychological conditions pertinent to the game. As such, the value of an athlete in a team is determined by his or her ability to compete under certain conditions. Furthermore, athletic competition, particularly at the elite level, is conducted in an atmosphere marked by high levels of tension and arousal as a result of the need to succeed in every competition [1, 2]. The term 'stress' is often used to denote a distressing mental state or illness characterized by subjective feelings of tension, apprehension, and worry [3]. In sports, this is referred to as precompetition tension or anxiety. In addition, tests have found that anxiety has a negative effect on these sports' outcomes [4]. As a result, in the modern athletic domain, the use of such therapeutic techniques to reduce the above stressors as well as increase success in sports has greatly advanced [2, 5]. The use of such therapeutic techniques is important for athletes because it helps them cope better during competitions.

© The Author(s), under exclusive license to Springer Nature Singapore Pte Ltd. 2021
R. Muazu Musa et al., *Machine Learning in Elite Volleyball*, SpringerBriefs
in Applied Sciences and Technology, https://doi.org/10.1007/978-981-16-3192-4_4

Psychological techniques have been documented to cover a wide variety of issues in the sport dimension. For example, some are designed to help with motivation initiation, control, and management [6], whilst others are designed to help with focus, self-confidence, and emotion regulation [7]. Volleyball, like many other non-inversion sports, is marked by a high degree of psychological demands as a result of the sport's fast-paced style. Therefore, recognizing the necessary psychological characteristics that can ensure success in the sport, especially at the elite level where the stakes are high, is a non-trivial task. The purpose of the present investigation is to determine the essential strategic competitional elements required in the elite men volleyball players.

Appraisal of the strategic competitional elements: The psychological components of the players in the sample were measured using the test performance strategies—competition scale (TOPS-SC), which was developed and validated by the previous researchers [8, 9]. During the competition, the TOPS-SC tested eight psychological coping mechanisms. These coping strategies constituted self-talk, activation, imagery, emotion control, automaticity, relaxation, goal setting as well as negative thinking. The players filled out this inventory, and their results were evaluated during the tournament as mentioned in the previous chapter. The performance of the players was also analysed in real time during the tournament. It is worth mentioning that players who are unable to play at least 70% of the time were not considered for the final analysis in this study.

4.2 Feature Selection

A number of feature selection technique is used in the present study, i.e. information gain (IG), gain ratio, ReliefF, and chi-square. These methods are used to extract the most important strategic competitional elements for volleyball players. It is worth noting that such a technique reduces overfitting, improves accuracy as well as reduces the training time as noises and redundant features are eliminated [10]. The players' strategic competitional scores, namely imagery, self-talk, activation, automaticity, emotional control, relaxation, negative thinking as well as goal setting, are evaluated.

4.3 Machine Learning-Based Regression Analysis

The present study investigates the efficacy of a number of machine learning-based regression models, namely support vector regressor (SVR), k-nearest neighbour regressor (kNNR) as well as decision tree regressor (DTR) in their ability to predict the performance of the players. The dataset that consists of 93 instances was split into a ratio of 70:30 for train and test, respectively [11, 12]. The models were developed and evaluated through Spyder IDE (Python 3.6). The default hyperparameters of the models were used based on the scikit-learn library. The models were evaluated by means of the coefficient of determination, R^2, as well as the mean squared error (MAE).

4.4 Results and Discussion

Table 4.1 tabulates the extracted vital competitional strategic variables influencing the positive performance of volleyball, as well as, the IG score foreach feature. It could be seen from the table that a total number of four competitional strategic variables are found to be the most significant features identified from the IG feature selection technique which are imagery, self-talk, activation, automaticity as well as emotional control.

The different models developed based on the significant features identified via the IG feature selection technique are shown in Fig. 4.1. It was demonstrated from the study that the SVR, kNNR, and DTR could achieve a test R^2 of 0.9999, 0.9759, and 0.9881, respectively. In addition, the test MAE recorded by the models is 0.0619, 1.1714, and 1.0, respectively, suggesting that the SVR model is able to predict the performance of the players based on the selected features. Figure 4.1 depicts the performance of the models against actual test data.

The strategic aspects of psychological preparation have been stated to be critical in enhancing athletic success in a variety of sports. Any psychological factors, such as fatigue and worry, have been shown to have a negative impact on the athletic success [3, 13, 14]. It has been demonstrated from the findings of the current investigation that certain psychological competitional variables, namely imagery, self-talk, activation, automaticity as well as emotional control, could be a major predictor of performance in the elite indoor men volleyball competition. These psychological elements could be characterized as an individual personal–psychological strategy that describes the quality of oneself in dealing with varying adversity during a sporting performance. Emotional control has been indicated to be an integral aspect in ensuring victory during sporting context [15]. The authors stated further that athletes' emotional states, such as happiness, enthusiasm,

Table 4.1 Feature selection techniques

Competition strategy variables	Information gain
Imagery	**0.121**
Self-talk	**0.095**
Activation	**0.070**
Automaticity	**0.042**
Emotional control	**0.033**
Relaxation	0.031
Negative thinking	0.029
Goal setting	0.027

An IG score of above 0.032 is considered to be significant for this case study

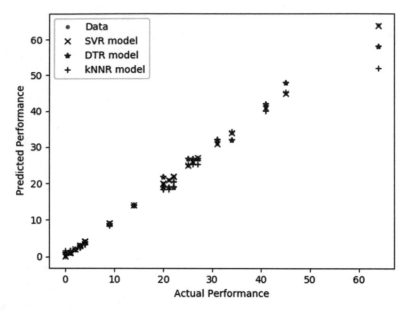

Fig. 4.1 Prediction performance of the different developed machine learning-based regression models

excitement, as well as sadness resulting from a humiliating loss, may influence athletic success and competition outcomes. Consequently, an athlete's ability to regulate his or her emotions can lead to improved performance during both training and competition.

4.5   Summary

The current study has identified the psychological competitional variables that are key in forecasting elite indoor volleyball players' success during tournaments. These combinations of competitional strategic variables that constituted imagery, self-talk, activation, automaticity as well as emotional control are shown to be pivotal in determining the performance of elite volleyball players during competition. Moreover, the findings further demonstrated that the linear-based SVR model is able to yield an accurate prediction of the performance of the players. It is also worth noting that the ability of an athlete to develop a personal strategy for coping with obstacles will help the athlete produce better results during competition in this sport to a greater extent. This point should be considered when creating a personal–psychological intervention for elite volleyball teams as well as an individual player.

References

1. Taha, Z., Musa, R.M., Abdullah, M.R., Maliki, A., Kosni, N.A., Mat-Rasid, S.M., Adnan, A., Juahir, H.: Supervised pattern recognition of archers' relative psychological coping skills as a component for a better archery performance. J. Fundam. Appl. Sci. **10**, 467–484 (2018)
2. Abdullah, M.R., Musa, R.M., Maliki, A.B.H.M.B., Kosni, N.A., Suppiah, P.K.: Role of psychological factors on the performance of elite soccer players. J. Phys. Educ. Sport. **16**, 170 (2016)
3. Musa, R.M., Abdullah, M.R., Juahir, H., Eswaramoorthi, V., Alias, N., Hashim, M.R., Alnamat, A.S.F.: An exploratory study of personality traits and psychological coping skills on archery performance. Indian J. Public Heal. Res. Dev. **10**, 630–635 (2019). https://doi.org/10.5958/0976-5506.2019.00572.2
4. John, S., Verma, S.K., Khanna, G.L.: The effect of mindfulness meditation on HPA-Axis in pre-competition stress in sports performance of elite shooters. Natl. J. Integr. Res. Med. **2**, 15–21 (2011)
5. Mohammadzadeh, H., Sami, S.: Psychological skills of elite and non-elite volleyball players. Ann. Appl. Sport Sci. **2**, 31–36 (2014)
6. Reiss, S., Wiltz, J., Sherman, M.: Trait motivational correlates of athleticism. Pers. Individ. Differ. **30**, 1139–1145 (2001)
7. Bois, J.E., Sarrazin, P.G., Southon, J., Boiché, J.C.S.: Psychological characteristics and their relation to performance in professional golfers. Sport Psychol. **23**, 252–270 (2009)
8. Abdullah, M.R., Maliki, A.B.H.M.: Reliability of test of performance strategies-competition scale (TOPS-CS) among… Man India. **96**, 5199–5207 (2016)
9. Thomas, P.R., Murphy, S.M., Hardy, L.E.W.: Test of performance strategies: Development and preliminary validation of a comprehensive measure of athletes' psychological skills. J. Sports Sci. **17**, 697–711 (1999)
10. Musa, R.M., Abdul Majeed, A.P.P., Musa, A., Abdullah, M.R., Kosni, N.A., Razman, M.A. M.: An information gain and hierarchical agglomerative clustering analysis in identifying key performance parameters in elite beach soccer. Presented at the (2021). https://doi.org/10.1007/978-981-15-7309-5_26
11. Taha, Z., Razman, M.A.M., Adnan, F.A., Abdul Ghani, A.S., Abdul Majeed, A.P.P., Musa, R.M., Sallehudin, M.F., Mukai, Y.: The identification of hunger behaviour of lates calcarifer through the integration of image processing technique and support vector machine. In: IOP Conference Series: Materials Science and Engineering (2018). https://doi.org/10.1088/1757-899X/319/1/012028
12. Musa, R.M., Majeed, A.P.P.A., Taha, Z., Abdullah, M.R., Maliki, A.B.H.M., Kosni, N.A.: The application of artificial neural network and k-nearest neighbour classification models in the scouting of high-performance archers from a selected fitness and motor skill performance parameters. Sci. Sports **34**, e241–e249 (2019)
13. Lazarus, R.S.: How emotions influence performance in competitive sports. Sport Psychol. **14**, 229–252 (2000)
14. Musa, R.M., Taha, Z., Majeed, A.P.P.A., Abdullah, M.R.: Psychological variables in ascertaining potential archers. In: Machine Learning in Sports, pp. 21–27. Springer (2019)
15. Jones, M.V.: Controlling emotions in sport. Sport Psychol. **17**, 471–486 (2003)

Chapter 5
Anthropometric Variables in the Identification of High-performance Volleyball Players

Abstract In this chapter, we delve into the importance of certain anthropometric- and age-related variables in the detection of high-performance players during an indoor volleyball tournament. It is demonstrated that high-performance players in this sport are essentially taller, relatively younger, as well as slightly heavier, whilst huge body size coupled with higher age is found to be associated with low performance in the sport of men indoor volleyball. Moreover, an ANN-based classification model was found to be effective in the identification of the performance classes of the players.

Keywords Body physique · Anthropometry · Indoor volleyball · High-performance players · ANN model

5.1 Overview

Individuals are normally classified based on their anatomical and body features, also known as body characteristics or anthropometry index. An individual's anthropometric index is made up of the composition, height, and shape of that person. Whilst the attribution of the test's dimensions can differ, anthropometry attributions such as height, weight, per cent body fat, and strength have been widely used by various studies to investigate the differences of such attributes with respect to a variety of athletic activities [1, 2]. Research also found that certain anthropometry characteristics, such as shape, size, and composition, could be used to predict a person's ability, resulting in the prediction of motor performance [3]. Furthermore, a positive substantial relationship between an individual's strength and anthropometry has previously been established in a variety of sports [4]. The study also found that stronger athletes have a specific body profile, which includes specific body size, shape, and indexes that distinguish them in the execution of motor performance. However, it is worth highlighting that certain body characteristics might not necessarily define or forecast athletic success in some sports.

R. Muazu Musa et al., *Machine Learning in Elite Volleyball*, SpringerBriefs in Applied Sciences and Technology, https://doi.org/10.1007/978-981-16-3192-4_5

Several studies have been conducted to determine the physical and physiological characteristics of volleyball players at both elite and amateur level [5–8]. The anthropometric profile and the somatotype characterization of elite volleyball players are not only important to determine the physical condition of the athletes, but also for their potential performance in relation to the function that they have during the volleyball game [9]. For instance, a recent study that compared the anthropometric measurement and body composition between junior basketball and volleyball players of the Serbian National League documented that the basketball and volleyball players were significantly taller and heavier than the other recreational players [5]. However, there was no significant difference between the body height and body weight of basketball and volleyball players. These findings contradict the previous research, in which significant differences were reported between the basketball and volleyball players in most of the anthropometry components examined. The basketball players were significantly higher in height, weight, body surface, body fat percentage, total body fat, fat free mass components as compared to volleyball players. Nonetheless, the volleyball players were significantly greater in body density as compared to basketball players [10].

The detailed information about the players' anthropometric characteristics, primarily their standing height, body weight, age, as well as their technical and tactical ability, may be used as reference values in the team selection and training process [11–13]. Furthermore, as volleyball players attain the peak level of performance age, these variables could guide the coach in team management. Age also provides a temporal reference for estimating how long it would take to reach optimum efficiency. However, a shred of scientific evidence is needed to ascertain the connection between the anthropometric variables and actual delivery of performance in the game by means of a randomized study protocol. The purpose of the present study is to examine the contribution of anthropometric variables towards the identification of high-performance volleyball players.

Anthropometry and skills assessment: Before the beginning of the tournament, the basic information of all the players concerning their standing height, weight as well as years of playing experience, was recorded. The standing height was collected in cm to the nearest 0.5 cm, whilst the weight was obtained in kg to the nearest 0.01 [14]. The technical and tactical ability of the players were determined using their performance in the execution of the skills that constitute ace, block, set, spike, fault, tap, dig as well as passing. The performances were notated in real time as detailed in the previous chapter. It should be noted that the players who are unable to play at least 70% of the time were not considered for the final analysis in this study.

5.2 Clustering

The hierarchical agglomerative cluster analysis (HACA) was utilized in this investigation. The performance of the players was initially clustered to obtain the performance class of the players. Subsequently, the anthropometric variables, namely standing height, body weight as well as age, were used to ascertain the differences of the initially developed clusters in the investigated anthropometric variables [15, 16].

5.3   Classification

In the present investigation, the efficacy of artificial neural networks (ANN) towards identifying the different performance classes clustered via HACA is evaluated. A two hidden layer ANN network topology is formulated, i.e. 3-X-X-1. The number of hidden neurons denoted as X is varied with 10, 50, 100, 150, 200, 250, and 300 on each hidden layer [17, 18]. The activation function is also varied between ReLU, logistic as well as hyperbolic tan (tanh). The optimization algorithm employed is Adam. It is worth noting that the learning rate is also varied between constant, invscaling as well as adaptive, whilst the iteration is set to 1000. The identification of the optimized hyperparameters is carried out via grid search technique through the five-fold cross-validation method on the test dataset. The dataset was initially split into a 70:30 ratio for training and testing, respectively. The model was developed on Spyder IDE running on Python 3.6 with its associated scikit-learn libraries. The performance of the model is deliberated via the classification accuracy as well as the confusion matrix.

5.4   Results and Discussion

Figure 5.1 projects the clusters allotted by the HACA algorithm in regard to the evaluated parameters in this investigation. It could be detected that three clusters were established based on the similarities in the characteristics of the variables assessed. The figures further illustrated a clear separation of the clusters which is non-trivial in facilitating the assignation of the identity to each of the clusters, viz. high physique performance (HPP), moderate physique performance (MPP) players as well as low physique performance players (LPP), respectively.

Table 5.1 tabulates the descriptive statistics of the variables investigated. The mean, as well as the standard deviation of each cluster, is projected. It could be observed that the clusters are different in the performance of the variables assessed which reflects that the high, moderate, and low physique performance players essentially distinguished by their peculiar attributes.

Figure 5.2 depicts the mean performance differences plots of the assigned clusters with respect to the anthropometric as well as the performance variables examined in the study. It could be observed from the boxplots that the HPP cluster is attributed to high skill performance, essentially taller, relatively younger, and slightly heavier, whilst the MPP is of moderate skill performance, a bit older,

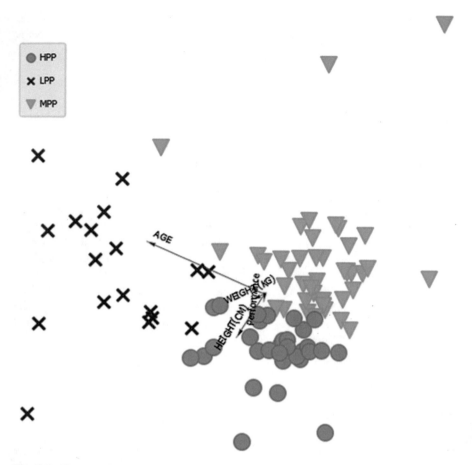

Fig. 5.1 Class membership allotted by the HACA clustering algorithm

Table 5.1 Physical characteristics and performance variations between the three clustered players

Anthropometric and performance variables	HPP		MPP		LPP	
	Mean	Std.D	Mean	Std.D	Mean	Std.D
Height	180.906	3.684	168.326	9.554	178.833	5.711
Weight	72.984	12.791	59.721	12.993	92.778	24.530
Age	21.031	2.265	21.814	3.239	32.278	4.240
Volleyball skill performance	20.469	17.246	12.605	14.151	11.333	9.864

shorter as well as less heavy. Interestingly, the LPP group is shown to be of comparatively low skill performance, slightly taller, heavier as well as the oldest amongst all the clusters.

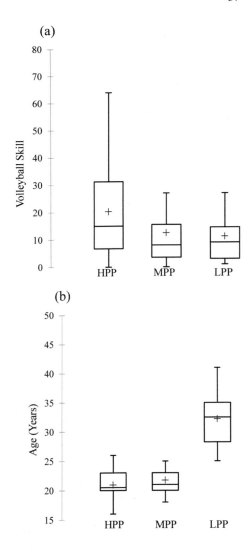

Fig. 5.2 Differences between the players on the anthropometry and skill performances. **a** volleyball skill, **b** age, **c** weight, **d** height

The efficacy of the ANN model in classifying the identified three performance classes, namely HPP, MPP and LPP, was investigated. From the grid search evaluation, it was found that the optimized number of hidden neurons for each hidden layers is 250 along with the logistic activation function and invscaling learning rate. The average classification accuracy attained was reported to be 0.8 ($\pm$0.23). The optimized model was then tested on the test dataset, and classification accuracy of 85.71% was obtained. Figure 5.3 depicts the confusion matrix of the test dataset.

Fig. 5.2 (continued)

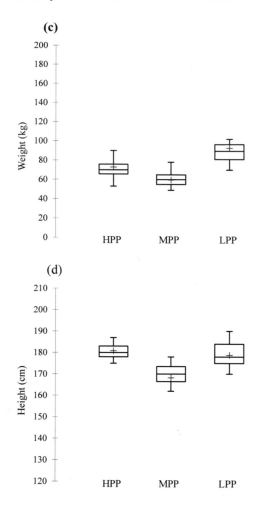

The findings from the present investigation demonstrated that there is a strong association between volleyball performance ability and physical characteristics in unification with the age. Physical attributes in conjunction with body composition are reported to be vital in the classification as well as differentiation of athletic performances [16, 18–20]. Research on men's volleyball has demonstrated that there is an increasing tendency to utilize tall players [21]. Growing evidence from different studies has shown that taller players were more likely to be used in modern indoor volleyball. For instance, a past study reported the 195 cm mean height of the elite volleyball players, whilst a relatively recent study documented a mean height of 200 cm, respectively [21, 22]. Comparing these studies demonstrated that volleyball players at the professional level have a 5 cm increased in height. It may then be postulated that the taller the player, the more likely that he would be selected against the other shorter players.

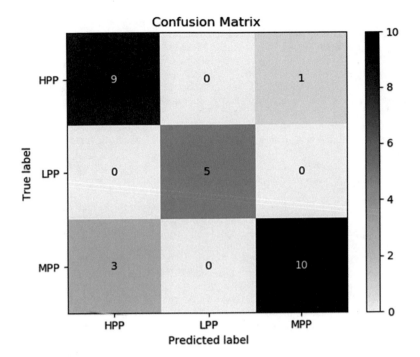

Fig. 5.3 Confusion matrix on the test dataset

5.5 Summary

It has been revealed from the findings of the current investigation that successful performance in an elite indoor volleyball tournament could be influenced by the players' body structure as well as age. It is demonstrated that high-performance players in this sport are essentially taller, relatively younger, as well as slightly heavier whilst huge body size coupled with higher age is found to be associated with low performance in the sport of indoor volleyball. Moreover, an ANN-based classification model was found to be effective in the identification of the performance classes of the players which is non-trivial in mapping out the high performance from a combination of anthropometric- and age-related variables.

References

1. Musa, R.M., Majeed, A.P.P.A., Kosni, N.A., Abdullah, M.R.: Identifying talent in sepak takraw via anthropometry indexes. In: Machine Learning in Team Sports, pp. 29–39. Springer (2020)
2. Musa, R.M., Taha, Z., Majeed, A.P.P.A., Abdullah, M.R.: Anthropometry correlation towards archery performance. In: Machine Learning in Sports, pp. 29–35. Springer (2019)

3. Ostojic, S.M., Mazic, S., Dikic, N.: Profiling in basketball: physical and physiological characteristics of elite players. J. Strength Cond. Res. **20**, 740–744 (2006)
4. Ball, T.E., Massey, B.H., Misner, J.E., Mckeown, B.C., Lohman, T.G.: The relative contribution of strength and physique to running and jumping performance of boys 7-11. J. Sports Med. Phys. Fitness **32**, 364–371 (1992)
5. Masanovic, B.: Comparative study of anthropometric measurement and body composition between junior basketball and volleyball players from Serbian national league. Sport Mont. **16**, 19–24 (2018)
6. Giannopoulos, N., Vagenas, G., Noutsos, K., Barzouka, K., Bergeles, N.: Somatotype, level of competition, and performance in attack in elite male volleyball. J. Hum. Kinet. **58**, 131 (2017)
7. Pastuszak, A., Busko, K., Kalka, E.: Somatotype and body composition of volleyball players and untrained female students-reference group for comparison in sport. Anthropol. Rev. **79**, 461 (2016)
8. Milić, M., Grgantov, Z., Chamari, K., Ardigò, L.P., Bianco, A., Padulo, J.: Anthropometric and physical characteristics allow differentiation of young female volleyball players according to playing position and level of expertise. Biol. Sport. **34**, 19 (2017)
9. Gualdi-Russo, E., Zaccagni, L.: Somatotype, role and performance in elite volleyball players. J. Sports Med. Phys. Fitness **41**, 256 (2001)
10. Gaurav, V., Singh, S.: Anthropometric characteristics, somatotyping and body composition of volleyball and basketball players. J. Phys. Educ. Sport Manag. **1**, 28–32 (2010)
11. Najmi, N., Abdullah, M.R., Juahir, H., Maliki, A., Musa, R.M., Mat-Rasid, S.M., Adnan, A., Kosni, N.A., Eswaramoorthi, V., Alias, N.: Comparison of body fat percentage and physical performance of male national senior and junior karate athletes. J. Fundam. Appl. Sci. **10**, 485–511 (2018)
12. Maliki, A.B.H.M., Abdullah, M.R., Juahir, H., Muhamad, W.S.A.W., Nasir, N.A.M., Musa, R.M., Mat-Rasid, S.M., Adnan, A., Kosni, N.A., Abdullah, F., Abdullah, N.A.S.: The role of anthropometric, growth and maturity index (AGaMI) influencing youth soccer relative performance. In: IOP Conference Series: Materials Science and Engineering. Institute of Physics Publishing (2018). https://doi.org/10.1088/1757-899X/342/1/012056
13. Abdullah, M.R., Maliki, A., Musa, R.M., Kosni, N.A., Juahir, H., Mohamed, S.B.: Identification and comparative analysis of essential performance indicators in two levels of soccer expertise. Int. J. Adv. Sci. Eng. Inf. Technol. **7**, 305–314 (2017)
14. Taha, Z., Haque, M., Musa, R.M., Abdullah, M.R., Maliki, A., Alias, N., Kosni, N.A.: Intelligent prediction of suitable physical characteristics toward archery performance using multivariate techniques. J. Glob. Pharm. Technol. **9**, 44–52 (2009)
15. Musa, R.M., Majeed, A.P.P.A., Kosni, N.A., Abdullah, M.R.: An overview of beach soccer, sepak takraw and the application of machine learning in team sports. Mach. Learn. Team Sport. 1–12 (2020)
16. Razali, M.R., Alias, N., Maliki, A., Musa, R.M., Kosni, L.A., Juahir, H.: Unsupervised pattern recognition of physical fitness related performance parameters among Terengganu youth female field hockey players. Int. J. Adv. Sci. Eng. Inf. Technol. **7**, 100–105 (2017)
17. Taha, Z., Razman, M.A.M., Adnan, F.A., Abdul Ghani, A.S., Abdul Majeed, A.P.P., Musa, R.M., Sallehudin, M.F., Mukai, Y.: The identification of hunger behaviour of lates calcarifer through the integration of image processing technique and support vector machine. In: IOP Conference Series: Materials Science and Engineering (2018). https://doi.org/10.1088/1757-899X/319/1/012028
18. Musa, R.M., Majeed, A.P.P.A., Taha, Z., Abdullah, M.R., Maliki, A.B.H.M., Kosni, N.A.: The application of artificial neural network and k-nearest neighbour classification models in the scouting of high-performance archers from a selected fitness and motor skill performance parameters. Sci. Sports **34**, e241–e249 (2019)
19. Abdullah, M.R., Musa, R.M., Kosni, N.A., Maliki, A., Haque, M.: Profiling and distinction of specific skills related performance and fitness level between senior and junior Malaysian youth soccer players. Int. J. Pharm. Res. **8**, 64–71 (2016)

20. Maliki, A.B.H.M., Abdullah, M.R., Juahir, H., Abdullah, F., Abdullah, N.A.S., Musa, R.M., Mat-Rasid, S.M., Adnan, A., Kosni, N.A., Muhamad, W.S.A.W., Nasir, N.A.M.: A multilateral modelling of youth soccer performance index (YSPI). IOP Conf. Ser. Mater. Sci. Eng. **342**, (2018). https://doi.org/10.1088/1757-899X/342/1/012057
21. Reilly, T., Secher, N., Snell, P., Williams, C.: Physiology of sports: an overview. Physiol. Sport. 465–485 (1990)
22. Stamm, R., Stamm, M., Jairus, A., Toop, R., Tuula, R., João, P.V.: Do height and weight play an important role in block and attack efficiency in high-level men's volleyball? Pap. Anthropol. **26**, 64–71 (2017)

Chapter 6
Performance Indicators Predicting Medallists and Non-medallists in Elite Men Volleyball Competition

Abstract In this chapter, we investigated the influence of technical and tactical performance indicators in determining the match outcome, i.e. medallists and non-medallists during an indoor volleyball competition. A set of performance indicators, namely the ability to block, spike as well as tap, are shown to be essential in determining the chances of a team to either earn or lose a medal. It has been shown that both technical and tactical skills are essential for ensuring success in the men elite indoor volleyball championship. Moreover, a stacking-based machine learning approach was found to be effective in the identification of the chances for winning or losing a medal during a volleyball competition.

Keywords Technical and tactical indicators · Indoor men volleyball · Stacking-based machine learning model · Medallists · Non-medallists

6.1 Overview

Performance appraisal is often used as a diagnostic instrument for evaluating the performance of teams or athletes [1–3]. In multiple sports, the use of performance analysis to measure and differentiate good and poor performance has been widely studied. Identifying tactical and technical approaches in the 2010 FIFA World Cup, establishing successful performance indicators in club-level Gaelic football, and developing a notational analysis framework in elite soccer and its resulting reliability are only a few examples of such studies [4–6]. Separate research assessed the most important success metrics in elite male soccer with regard to different playing positions [7]. In a more recent study, the effect of technical and tactical performance parameters in ascertaining the final outcome of a match (successful or unsuccessful) in an elite beach soccer tournament was investigated [8]. The authors inferred that certain performance parameters could potentially discriminate the outcome of a match in the sport.

Most of the volleyball-related research in evaluating technical and tactical performance has been implemented with senior players, including in indoor volleyball

[9, 10] as well as beach volleyball [11–13]. Some conclusions were made from the findings of the studies that could be summarized as follows: the attack (points and opponent errors) and services are the crucial performance parameters that could predict success in the senior male beach volleyball. On the other hand, higher reception ability, as well as attack effectiveness, is the major technical and tactical elements discriminating the senior and junior male players indoor volleyball [14, 15]. Moreover, data from a relatively recent study demonstrated that the winning accumulated more points during counter-attack particularly from forced and unforced errors of the opponents' block, serves as well as other related events [16].

It could be observed that adequate attentions have been given to studying the success metrics that could differentiate between winning and losing teams in both beach as well as indoor volleyball at a different level of expertise. However, no data is yet reported on the relevant performance indicators that could predict the probability of a team winning or losing a medal, particularly in an elite congested fixtures tournament format. Hence, the purpose of the present study is to investigate the essential performance indicators that could predict medallist and non-medallist in an elite men indoor volleyball competition.

6.2 Performance Indicators Development

For assessing the success of each individual player on the squad, a total of eight technical and tactical performance criteria were considered. The players' performance was analysed in real time using the performance criteria of ace, block, set, spike, fault, tap, dig, and passing. It is worth mentioning that the above success metrics were chosen based on their importance to the game of volleyball, as shown by previous research [11, 12, 17]. The StatWatch system was the medium employed for analysing the performance of the players and teams in compliance with the guidelines previously stated by the previous investigators [6]. For the detailed procedures of the analysis as well as the reliability technique, the readers could refer to the previous chapter (Chap. 1).

6.3 Classification

In the present study, four baseline machine learning models are evaluated, namely logistic regression (LR), random forest (RF), decision tree (Tree), and Naïve Bayes (NB), in identifying medallist and non-medallists. In addition, an ensemble technique, i.e. stacking is appraised. Stacking exploits the strength of the different baseline models to allow for a better prediction. The meta-model (second layer) used in the present study is the LR model. Figure 6.1 depicts the proposed stacked methodology. The dataset was split into a 70:30 ratio for testing and training [18, 19]. From the training dataset, the individual models, as well as the stacked model,

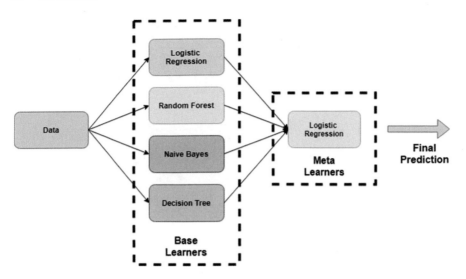

Fig. 6.1 Stacking Approach

are evaluated through repeated stratified k-fold technique in which the dataset is split into three-fold and repeated five times. The models are evaluated based on the classification accuracy performance metrics. The evaluation was carried out on a Python 3.6 powered IDE, Spyder.

6.4   Results and Discussion

Table 6.1 reveals the performance indicators distinguishing the medallist and non-medallist in the elite indoor volleyball tournament. It could be seen that out of the eight initially investigated performance indicators, only three were found to be statistically significant, viz. block, spike as well as tap $p \leq 0.05$. These perfor-mance indicators could be categorized as a combination of both technical and tactical skills. It could be postulated that a combination of both technical and tactical strategies is essential for ensuring success in the men elite indoor volleyball championship.

Figure 6.2 portrays the mean performance differences plots of medallist and non-medallist in the elite indoor volleyball tournament. It could be observed from the boxplots that the medallists are ascribed with high performance in the evalu-ated performance indicators. Remarkably, the medallists recorded a comparatively high technical as well as tactical skill performance as opposed to the non-medallists.

Table 6.1 Technical and tactical performance indicators differentiating the medallist and non-medallist

Performance indicators	Team achievement	Mean	Std. deviation	P-value (MWU)
Ace	Non-medallist	8.842	9.161	0.101
	Medallists	12.000	2.944	
Block	Non-medallists	6.684	4.243	0.050*
	Medallists	11.250	2.630	
Set	Non-medallists	4.368	5.580	0.204
	Medallists	8.250	6.850	
Spike	Non-medallist	33.526	23.296	0.039*
	Medallists	63.750	22.292	
Fault	Non-medallists	37.526	18.470	0.155
	Medallists	52.000	15.470	
Tap	Non-medallists	5.895	4.254	0.049*
	Medallists	11.000	4.243	
Dig	Non-medallists	0.316	0.749	0.955
	Medallists	0.250	0.500	
Passing	Non-medallists	3.789	3.780	0.479
	Medallists	8.250	9.359	

$*P \leq 0.05$

Figure 6.3 illustrates the boxplot on the performance of the models evaluated on the training dataset. It could be seen that the average classification accuracy of the LR, RF, tree, NB and stacking is 0.518 ($\pm$0.206), 0.673 ($\pm$0.128), 0.556 ($\pm$0.153), 0.760 ($\pm$0.089) and 0.811 ($\pm$0.016), respectively. It is evident that the ensemble model yielded a higher classification accuracy. The stacked model is able to achieve a CA of 85.71% on the test dataset, suggesting the efficacy of the proposed model in classifying medallist and non-medallist.

The overall findings from this study demonstrated that a combination of certain performance indicators could potentially serve as a distinctive marker between the medallist and non-medallist in the elite indoor volleyball tournament. The performance indicators specifically the ability to block, spike as well as tap are shown to be essential in determining the chances of a team to either earn a medal or otherwise. It should be noted that these performance indicators could be characterized as a combination of both technical and tactical skills. It could then be suggested that a combination of both technical and tactical strategies is essential for ensuring success in the men elite indoor volleyball championship.

Fig. 6.2 Significant
performance indicators
differentiating medallist from
non-medallists in an indoor
men volleyball tournament
a tap, **b** spike, **c** blocks

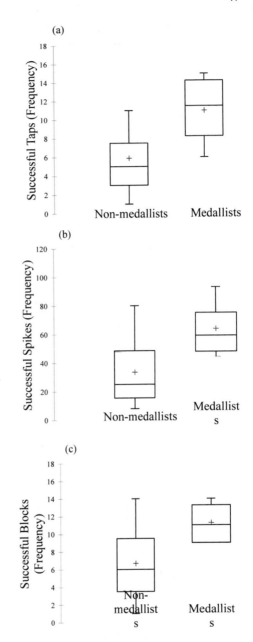

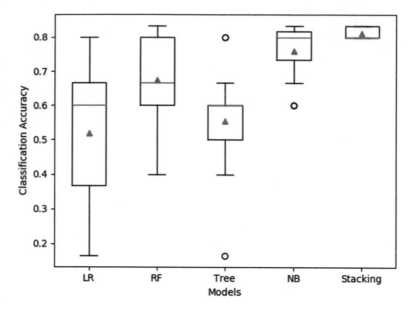

Fig. 6.3 Performance of the individual models against the ensemble 'stacking' model

6.5 Summary

The present investigation has established the key performance indicators that distinguish between medallists and non-medallists in men elite indoor volleyball tournament. A set of performance indicators, namely the ability to block, spike as well as tap, are shown to be essential in determining the chances of a team to either earn or lose a medal. It has been shown that both technical and tactical skills are essential for ensuring success in the men elite indoor volleyball championship. Thus, the combination of the said performance indicators should be considered when designing a notational analysis system in elite men indoor volleyball tournaments.

References

1. A Gomez, M., Lago-Peñas, C., Viaño, J., González-Garcia, I.: Effects of game location, team quality and final outcome on game-related statistics in professional handball close games. Kinesiol. Int. J. Fundam. Appl. Kinesiol. **46**, 249–257 (2014)
2. Arnason, A., Sigurdsson, S.B., Gudmundsson, A., Holme, I., Engebretsen, L., Bahr, R.: Physical fitness, injuries, and team performance in soccer. Med. Sci. Sport. Exerc. **36**, 278–285 (2004)
3. Muazu Musa, R. P. P., Abdul Majeed, A., Abdullah, M.R., Ab. Nasir, A.F., Arif Hassan, M. H., Mohd Razman, M.A.: Technical and tactical performance indicators discriminating winning and losing team in elite Asian beach soccer tournament. PLoS One **14**, e0219138 (2019). https://doi.org/10.1371/journal.pone.0219138

4. Casal, C.A., Andujar, M.Á., Losada, J.L., Ardá, T., Maneiro, R.: Identification of defensive performance factors in the 2010 FIFA World Cup South Africa. Sports **4**, 54 (2016)
5. McGuigan, K., Hughes, M., Martin, D.: Performance indicators in club level Gaelic football. Int. J. Perform. Anal. Sport. **18**, 780–795 (2018). https://doi.org/10.1080/24748668.2018.1517291
6. Abdullah, M.R., Musa, R.M., Maliki, A.B.H.M., Kosni, N.A., Suppiah, P.K.: Development of tablet application based notational analysis system and the establishment of its reliability in soccer. J. Phys. Educ. Sport. **16**, 951–956 (2016). https://doi.org/10.7752/jpes.2016.03150
7. Hughes, M.D., Caudrelier, T., James, N., Redwood-Brown, A., Donnelly, I., Kirkbride, A., Duschesne, C.: Moneyball and soccer-an analysis of the key performance indicators of elite male soccer players by position (2012)
8. Musa, R.M., Majeed, A.P.P.A., Kosni, N.A., Abdullah, M.R.: Technical and tactical performance indicators determining successful and unsuccessful team in elite beach soccer. In: Machine Learning in Team Sports. pp. 21–28. Springer (2020)
9. Afonso, J., Mesquita, I.: Determinants of block cohesiveness and attack efficacy in high-level women's volleyball. Eur. J. Sport Sci. **11**, 69–75 (2011)
10. Drikos, S., Vagenas, G.: Multivariate assessment of selected performance indicators in relation to the type and result of a typical set in men's elite volleyball. Int. J. Perform. Anal. Sport. **11**, 85–95 (2011)
11. George, G., Panagiotis, Z.: Statistical analysis of men's FIVB beach volleyball team performance. Int. J. Perform. Anal. Sport. **8**, 31–43 (2008)
12. Giatsis, G., Tzetzis, G.: Comparison of performance for winning and losing beach volleyball teams on different court dimensions. Int. J. Perform. Anal. Sport. **3**, 65–74 (2003)
13. Michalopoulou, M., Papadimitriou, K., Lignos, N., Taxildaris, K., Antoniou, P.: Computer analysis of the technical and tactical effectiveness in Greek Beach Volleyball. Int. J. Perform. Anal. Sport. **5**, 41–50 (2005)
14. García-Alcaraz, A., Palao Andrés, J.M., Ortega, E.: Perfil de rendimiento técnico-táctico de la recepción en función de la categoría de competición en voleibol masculino (2014)
15. García-de-Alcaraz, A., Ortega, E., Palao, J.M.: Effect of age group on male volleyball players' technical-tactical performance profile for the spike. Int. J. Perform. Anal. Sport. **15**, 668–686 (2015)
16. Medeiros, A.I.A., Marcelino, R., Mesquita, I.M., Palao, J.M.: Performance differences between winning and losing under-19, under-21 and senior teams in men's beach volleyball. Int. J. Perform. Anal. Sport. **17**, 96–108 (2017)
17. Giddens, S., Giddens, O.: Volleyball: Rules, Tips, Strategy, and Safety. The Rosen Publishing Group (2005)
18. Taha, Z., Razman, M.A.M., Adnan, F.A., Abdul Ghani, A.S., Abdul Majeed, A.P.P., Musa, R.M., Sallehudin, M.F., Mukai, Y.: The identification of hunger behaviour of Lates Calcarifer through the integration of image processing technique and support vector machine. In: IOP Conference Series: Materials Science and Engineering (2018). https://doi.org/10.1088/1757-899X/319/1/012028
19. Musa, R.M., Majeed, A.P.P.A., Taha, Z., Abdullah, M.R., Maliki, A.B.H.M., Kosni, N.A.: The application of artificial neural network and k-nearest neighbour classification models in the scouting of high-performance archers from a selected fitness and motor skill performance parameters. Sci. Sports **34**, e241–e249 (2019)

Chapter 7
Summary, Conclusion, and Future Direction

Abstract This chapter shall provide the summary of the analysis carried out, draws the conclusion of the findings as well as provides some insight into the future direction of the study.

Keywords Machine learning · Human performance · Indoor elite volleyball

7.1 Summary

Volleyball has metamorphosed into a high-degree strength sport and is considered as one of the most popular open-skilled-based team sports. The characteristic of the sport as an open-based skill required the players to possess a high level of both perceptual skills as well physical prowess to cope with the externally and internally induced pace of the sport. The players are expected to quickly make decisions and align their skills to the changing or otherwise unpredictable competitive environment. With a playing court of 900 square feet and having six players on each side of the net playing the game at a time, necessitates the players to attenuate the opposing players from hitting the ball across their side of the court, coordinate team movement via reading the game, reacting as well as moving rapidly whilst the ball is in play. Hence, the players may not have control over what will happen during the game situation. However, to ensure success during the game and to effectively execute the skills, the possession of certain performance elements becomes imperative. In the present brief, we employed various ML algorithms to evaluate the training and competitional strategies towards volleyball performances as well as identify high-performance volleyball players. Several psychological element strategies coupled with human performance parameters were explored in view to ascertain their impact on performance in men indoor elite volleyball competitions.

7.2   Conclusion

It has been demonstrated from the findings of the current investigation that several performance elements are essential in ensuring success in the sport of volleyball during both training and competition. It is shown that some sets of psychological training elements that included attentional control, imagery, relaxation, activation, goal setting, emotional control, automaticity as well as self-talk are non-trivial in predicting the performance of the elite volleyball players. Similarly, numerous competitional coping elements that constituted imagery, self-talk, activation, automaticity as well as emotional control are shown to be pivotal in determining the performance of elite volleyball players during competition. On the other hand, the findings from the present brief demonstrated that there is a strong association between volleyball performance ability and physical characteristics in unification with the age of the players. It is demonstrated that high-performance players in this sport are essentially taller, relatively younger, as well as slightly heavier whilst huge body size coupled with higher age is found to be associated with low performance in the sport of men indoor volleyball championship. It is also worth mentioning that certain technical and tactical performance indicators could potentially determine the match outcome, i.e. medallists and non-medallists during competition in this sport. These performance indicators were found to be the ability to block, spike as well as tap that are essential in determining the chances of a team to either earn or lose a medal during a tournament. It is also worth concluding that the application of machine learning models is pivotal in solving both classification and regression problems associated with volleyball performance-related data used in the current investigation. The findings from the present brief illustrated the key performance indicators as well as human performance parameters that could be used in the future evaluation of team performance as well as the identification of high-performance players in the elite volleyball tournament. It is believed that the findings herein could be vital to the coaches, club managers, talent identification experts, performance analysts as well as other important stakeholders in the evaluation of performance to foster improvement in this sport.

7.3   Future Direction

Machine learning (ML) techniques have emerged as one of the essential twenty first-century skills sequels to its contribution towards solving a real-world problem in a variety of fields. The popularity of the ML in the current century is attributed to its ability in handling big datasets, mitigating the nonlinearity issues in a dataset as well as making a robust prediction that could be used to seek information about future uncertainties. Hitherto, the data obtained in a sporting domain during both training and competitions, is increasingly becoming complex, hence making it difficult to decipher using the traditional statistical analysis. Thus, the application of ML in evaluating sporting performances is a non-trivial task. It is, therefore, envisage that the technique employed in the current brief should be extended towards other sports and physical activity-related domain.